Hydraulic Canals

Hydraulic Canals

Design, construction, regulation and maintenance

José Liria Montañés

CRC Press
Taylor & Francis Group
Boca Raton London New York

CRC Press is an imprint of the
Taylor & Francis Group, an **informa** business
A TAYLOR & FRANCIS BOOK

CRC Press
Taylor & Francis Group
6000 Broken Sound Parkway NW, Suite 300
Boca Raton, FL 33487-2742

First issued in paperback 2019

© 2006 by Taylor & Francis Group, LLC
CRC Press is an imprint of Taylor & Francis Group, an Informa business

Typeset in Sabon by
Integra Software Services Pvt. Ltd, Pondicherry, India

No claim to original U.S. Government works

ISBN-13: 978-0-415-36211-5 (hbk)
ISBN-13: 978-0-367-44640-6 (pbk)

British Library Cataloguing in Publication Data
A catalogue record for this book is available from the British Library

Library of Congress Cataloging in Publication Data
Liria Montañés, José.
  Hydraulic canals : design, construction, regulation
  and maintenance / José Liria Montañés.—1st ed.
    p. cm.
  ISBN 0–415–36211–3 (hardback)
  1. Canals—Design and construction. 2. Canals—Maintenance
  and repair. I. Title.
  TC745.L573 2005
  627′.13—dc22
                                              2005013676

Visit the Taylor & Francis Web site at
http://www.taylorandfrancis.com

and the CRC Press Web site at
http://www.crcpress.com

Contents

Figures

Preface to the Spanish edition

Although the section on canals is a very important topic within the subject of Hydraulic Engineering, the technical literature that has been written is aimed more directly at other work, such as dams, hydroelectric plants, piping, drinking water supply to towns, etc.

There are many excellent books on canals from the theoretical hydraulic performance point of view, but publications from the engineering design, construction and handling points of view are few and far between.

Many engineers, when starting out on their professional careers within this specialty, needed – and could not find – a text that resolved their doubts about details, that gave hints as to how to obtain complementary documentation and, above all, that gave an overall view of the subject.

By dint of asking first some and then other people, by searching ceaselessly for the most diverse bibliography and having made countless mistakes, it seemed to us that the writing of this book could be very useful.

It is aimed at hydraulic engineers and therefore it involves a large amount of theoretical knowledge on this science, without which a large part of its usefulness would be lost. Notwithstanding this, when it seemed to us to be useful to the reader for references to the theoretical basis on which a point is founded to be made, we have not hesitated in quoting them, even developing some mathematical demonstrations which, as they are not very usual, could run the risk of becoming forgotten. Likewise, the first chapter included serves as a reminder of the most important points of Hydraulics to lay out the background for all the problems studied in the book.

We have tried to fill the book with the experience acquired in 45 years of professional work, carried out in many different countries, sometimes through participative experience and at other times through observation. Also, everything that has been learnt from study, from conversations with other technicians and – a very important point – from the questions that our countless students have asked us is included. To all those who have helped us to learn something (engineers, students, builders, water users, etc.) we would like to express our deepest gratitude. We believe that, as in certain branches of Civil Engineering (for example in the calculation of Structures) there are physics–mathematical theories that lead to calculations that are virtually exact, on the subject of Canals there is still a gap in the scientific study that should be filled in to be used as a basis to the calculation and design of certain building elements. We must fight to replace a recipe of methodologies, which is passed on from one to another, with a theoretical basis that is scientifically proven. On this aspect, we have spared no effort in analysing and justifying certain design criteria, and therefore we believe that this book has some original points, which we have had the opportunity to check out in practical experience.

Preface

The reasons for writing this book, which were given in the preface to the Spanish edition, continue to be perfectly valid. In fact, we believe that it has been of great use to many Spanish and Latin American engineers and, because of similarity in languages, also to quite a number of Portuguese-speaking engineers. However, there are many hydraulic engineers throughout the world who do not understand Spanish and for this reason we felt it necessary to produce this English edition.

The time that has elapsed between the two editions (some three years) has allowed the opinions of many experts to be received and this enabled us to continue analysing what we could call "Canal Engineering". We feel that there is a significant demand in the engineering field for information on red-hot issues, such as canal operations, their rehabilitation and maintenance. For this reason we have enlarged the chapters referring to these subjects. However, we have not been able to escape from the very evident fact that canal design and construction is a much more developed technical theme that has led to a large part of the book being dedicated to it.

We want to point out that the sole purpose of this book is to be of help to engineers working on Canals. But the possible problems are so many and so diverse, and may be so complicated that the book cannot act as a substitute for the knowledge and good judgement of the engineer who has to take decisions under his own responsibility, without placing it upon the author of this book.

Looking back on our long professional life, we can see such a large number of people who, consciously or not, have helped us in acquiring the knowledge included in this book that we find it impossible to name them all, without running the risk of forgetting some. We would, therefore, like to take this opportunity of offering all of them our most sincere gratitude.

Professor José Liria Montañés has a Ph.D. Degree in Civil Engineering and is also a retired university professor.

His professional career has almost always been involved in hydraulic works and very especially dedicated to canals employed for irrigations purposes, as well as hydroelectric installations and drinking water supplies. He has worked on the design, construction and operation of canals for the Hydrographic Confederation of the Guadiana River for many years and as Project and Works Manager for the Isabel II Canal, Madrid.

He has also been deeply involved as a consulting engineer, not only in Spain, but also in many other countries, such as Venezuela, Brazil, Ecuador, Uruguay, Argentina and Peru, etc. This experience has provided him with a wider, more generalized view of the problems that can arise in canals.

He has written several books on the piping of water and others related to canals, at the moment only in Spanish, such as *Town Water Supply Networks* and *Modelling Special Works in Small Canals*.

His double activities both as a professional and in teaching have meant that he has had to analyse canals, not only from his own personal experience, but also from the point of view of the reasoned justification for the solutions to be adopted.

He is currently the Secretary for Spanish Committee on Irrigation and Drainage.

Chapter 1

Hydraulic operation of a canal

1.1 Basic general principles for its calculation

A canal is nothing but an open artificial channel used to carry water by means of a man-made river.

The water circulating inside a canal runs at a certain speed, producing mechanical forces between the water and the walls and the bottom of the canal due to its rubbing against them.

The influence is mutual; on the one hand, the wet inner surface of the canal rubbing on the water tends to slow down its movement. On the other hand, the water tends to erode the walls and bottom of the canal, and the energy of the moving water is capable of carrying solid particles that have been either broken off from the canal itself or entered the canal otherwise.

The force of the water on the walls and bottom of the canal is a shear force and is usually called an erosive force (because of its ability to erode the canal by pulling off particles) or tractive (because of its ability to carry the said particles). By the principle of action equal to reaction, the force of the wet perimeter of the canal rubbing on the water is equal to the former and is therefore called by the same names.

1.2 Calculation of average traction

From the theoretical viewpoint and especially the engineering point of view, it is very important to be able to have an idea, albeit only approximate, of what the traction is in a given section of a canal. Suffice it to think that if the resistance to erosion (which can be measured experimentally for the various types of soil) is greater than the erosive force, there will be no erosion. The opposite is also true.

As we will see, traction is not constant throughout the entire inner perimeter of a cross section of a canal; there are points in a single section in which it is much less than in others. Nevertheless, in the first attempt, let us calculate average traction along the inner perimeter of a single section, i.e. that uniform rate of traction that produces an effect on the water close enough to that which would be produced by the real rate of traction.

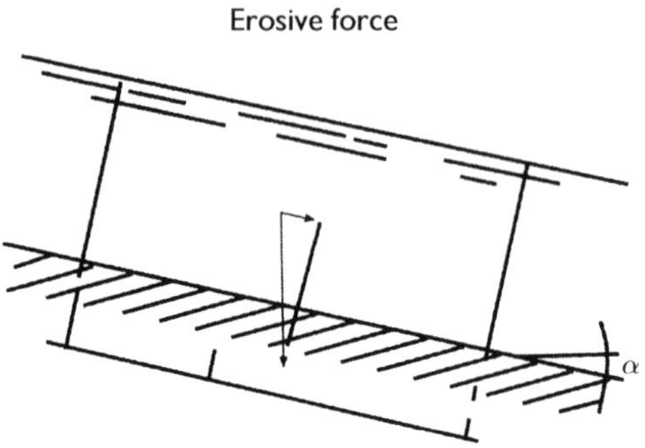

Erosive force

A = Cross-sectional area
P = Wetted perimeter
R = Hydraulic radius = A/P
F = Force moving water
s = Gradient
γ = Specific weight of the liquid
t = Tractive or shear force

Figure 1.1 Average traction.

Figure 1.1 is a schematic representation of a section of a canal with a sloping floor in relation to the horizontal plane, measured by angle *a*, the trigonometric tangent of which we shall call *s*.

For our estimate we shall consider a stretch of moving water, of length *L*, which is actually prismatic in shape.

Active thereon are forces that tend to compel it to slide in a downhill direction (the component of its weight parallel to the canal floor) and other forces that tend to brake its movement (the force of the terrain rubbing on the water, which is equal to the traction).

If we call *A* the area of the cross section, *P* the wet perimeter of this cross section, and define the hydraulic radius *R* as the quotient *A/P*, the volume of the prism formed by the section of the canal is $A \cdot L$.

We note that the hydraulic radius has the dimensions of a length.

Calling γ the specific weight of the liquid circulating through the canal (for us this will always be water, i.e. $\gamma = 1000 \, \text{kg/m}^3$, although in a general calculation it may be that corresponding to any liquid), the weight of the prism will be $A \cdot L \cdot \gamma$.

That weight acts as a vertical force, which is divided into two, one perpendicular to the bottom or floor of the canal (and which is resisted thereby), and the other parallel to the floor, whose value is $A \cdot L \cdot \gamma \cdot \sin a$. This is precisely the force that tends to move the section of water in a direction parallel to the slope of the floor, i.e. downhill.

If we call t the average rate of shear force by surface unit, which we suppose is constant throughout the entire perimeter of the canal section, for purposes of simplification, the force that opposes the movement of the water, as a result of the terrain's rubbing against the water, will be $t \cdot P \cdot L$.

The section of water considered may be moving at an increasingly faster rate of speed, an increasingly slower rate of speed, or at a uniform rate. It all depends on the relation between the forces that tend to move it or slow it down.

The case most frequently studied in canals, at least in an initial approach, is that of uniform movement. This we define as that in which the water throughout the canal neither speeds up nor slows down; therefore, the molecules in the stretch travel at the same rate of speed and the water prism retains the same form throughout. The surface level of the water must be parallel to the floor, in order for the section being considered to adopt the same condition and form as that of the following section, which space will be ocuppied by the next section.

In order for this constant rate of speed to occur, the difference between the forces that cause the water to move and those that slow it down must be zero, so that there will be no increase in the rate of speed (recall that force is equal to mass by acceleration), or, stated otherwise, $A \cdot L \cdot \gamma \cdot \sin a = t \cdot P \cdot L$.

The grade in canals is usually very slight (for example, 10–20 cm of drop in level per kilometre in very large canals, or from 80–100 cm/km in the most frequent size canals). For such small values the angle in radians is known to be the same as its trigonometric tangent or its sines; therefore, we may say $\sin a = s$.

By substituting this value, and by simplifying and recalling our definition of hydraulic radius, the former equality may be expressed as $R \cdot y \cdot s = t$, which gives us an initial value for average traction.

As stated earlier, we generally only manage water. Therefore, upon substituting the value for γ, we will have $t = 1000 \cdot R \cdot s$, a very useful formula and one of great importance, as we will see further on.

This means that if, for example, we have a canal with a 20 cm/km slope, i.e. a gradient of 2/10 000 and a hydraulic radius of 3 m (a large canal is involved), the tractive force will be

$$1000 \, \text{kg/m}^3 \times 3 \, \text{m} \times 0.0002 = 0.6 \, \text{kg/m}^2$$

If the canal had a gradient of 90 cm/km and a hydraulic radius of 1.10 m, the traction would be:

$$1000 \, \text{kg/m}^3 \times 1.10 \, \text{m} \times 0.0009 = 0.99 \, \text{kg/m}^2$$

If the terrain has a resistance of over $0.6 \, \text{kg/m}^2$ in the first instance or $0.99 \, \text{kg/m}^2$ in the second, there will be no erosion; otherwise, there will be.

1.3   Application to the calculation of the rate of flow circulating through a canal

The interrelationship between the erosive force of the water on the one hand and the force of the land rubbing on the water on the other is the basis for our ability to calculate traction. Yet equally it enables us to calculate the velocity of the water that produces such force.

The study of hydraulic theory teaches that head losses (or losses in height of the hydraulic grade line) of the water are different depending on whether the water movement is laminar or turbulent. In the first instance, the head loss of the water is proportional to the speed. In the second, it is proportional to the square of the velocity.

A typical example of laminar movement is that of underground water approaching a well, or that of slow moving water in very small diameter drip irrigation tubes, whereas water that runs in canals and ditches is almost always turbulent.

Therefore, the equation that will give us the tractive force in a canal in function of speed v will be:

$$T = k \cdot v^2$$

Substituting therein the value obtained for tractive force, we will have:

$$T = c \cdot R \cdot s = k \cdot v^2$$

One deduces thereby that

$$V = c' \cdot (R \cdot s)^{1/2}$$

which enables us to calculate the speed in a canal in terms of its hydraulic radius and gradient, and is the well-known Chezy formula, in which c' is a coefficient that varies essentially depending on how rough the canal perimeter is, which we shall simply call c from now on.

But the value c used in estimating the head loss caused by rubbing in the canal also depends, as can be easily understood, on the shape of the cross section of the canal; therefore, giving it a fixed value, dependent only on the roughness of the walls, is merely a first approach.

In order to better attune the formula of speed and bearing this concept in mind, attempts have been made through laboratory studies and studies of actual canal cases to obtain values of the c value in the Chezy formula that somehow includes the shape of the canal, in addition to the roughness of the walls.

It is easily apprehended that a simple parameter that varies with the shape of the section, and to a certain extent that helps classify the properties of its different basic forms, is the hydraulic radius. For very wide canals with little depth, the hydraulic radius is quite similar to the depth y. If it is a matter of a rectangular section the width of which is double the depth, the hydraulic radius is 0.5y, a value that drops to y/3 for the square cross section. As a result, the hydraulic radius is a coefficient multiplied by the depth. When the canal becomes very deep and narrow, such coefficient tends to drop to zero.

All canal calculation formulas are basically the same, for they are all derived from the Chezy formula, substituting a value based on the hydraulic radius for the coefficient of roughness c.

Manning proposed substituting the following formula for the value c in the Chezy formula:

$$c = 1/n \cdot R^{0.16}$$

which gave way to the formula bearing his name, which is:

$$V = 1/n \cdot R^{2/3} \cdot s^{1/2}$$

in which n is a coefficient of roughness.

Bazin proposed the following:

$$c = \frac{87}{(1 + \Gamma/\sqrt{R})}$$

This resulted in the formula given below that bears his name, in which Γ is the coefficient of roughness:

$$v = \frac{87 \cdot \sqrt{R} \cdot \sqrt{s}}{(1 + \Gamma/\sqrt{R})}$$

In the same way, many other researchers and engineers (e.g. Kutter, Pavloski, etc.) proposed diverse formulas, based on approximate values for c in the Chezy formula according to diverse expressions based on R.

The classic values of the coefficients of roughness in the prior formulas are the following (46):

Bazin's Formula

	Coefficient of roughness (Γ)
Plaster or layer of very good cement. Brushed plank. Sheet plates without rivets, etc. All with straight lines and clean water	0.06
The same case as above with rough water and medium-radius curves	0.10
Cement layer less refined than in the prior case Brushed plank but joined. Riveted steel plate. Lining made with cutstone. Not wide curves. Clean water	0.16

(Continued)

	Coefficient of roughness (Γ)
Cement plaster careless made. Plank not brushed, with joints. Unlined canals with excellent construction and maintenance, with bottom and slopes without vegetation. Wide Curves. Bottom with scarce sediment	0.36
Gun cement, etc.	0.40
Concrete walls with no outer layer, with protruding joints. Slime or moss on the walls and bottom. Tortuous design. Covering with regular stone work	0.46
Uniform sections on land or covered with gravel, no vegetation, and wide curves. Irregular stone work; smooth floor with mud deposit	0.85
Well preserved unlined canals; smooth walls or flat floor with low vegetation on the walls. Old stone work lining and muddy bottom	1.00
Unlined canals, with low weeds on the bottom and walls. Rivers and streams with irregular course but without vegetation	1.30
Unlined canals, abundant vegetation, erosion and irregular deposits of gravel. Neglected maintenance	1.75
Poorly maintained canals, with unconnected banks; canals with vegetation making up large part of the section	2.30

As a reference, we note that the Standards for the Design and Construction of small Irrigation Canals, published by the then Spanish Ministry of Public Works, stipulated that the design of concrete irrigation canals built in situ and manually should be done with Bazin's formula and 0.30 coefficient of roughness.

For Manning's formula, and according to the same source (45), the following values may be used:

Manning's Formula

	Coefficient of roughness (n)
Very smooth layers of plaster	0.010
Ordinary layers of plaster, coarse wood	0.011
Packed-down concrete, good stone work	0.012
Regular concrete, masonry	0.013
Used metallic ducts, regular stone work	0.017
Fine gravel, stone work in poor condition	0.020
Stone-free natural canals and rivers	0.025
Natural canals and rivers with stone and grass	0.030

It should be pointed out that some books and authors, in using Manning's formula, place the roughness coefficient n not in the denominator but in the numerator. Actually we are then in Strickler's formula and the coefficients to be used are logically the inverse values of those given for Manning.

Once the coefficient of either Bazin or Manning has been chosen, if one wants to determine the equivalent coefficient in Chezy's formula, one need only introduce the coefficient selected in the matching formula among those mentioned and that links Chezy's coefficient with that of Bazin or Manning. The value that multiplies to

$$(R \cdot s)^{1/2}$$

is that of the coefficient of Chezy.

1.4 Localized head losses

Localized head losses at given points are caused by water turbulence at more or less manifest obstacles scattered throughout the canal.

As explained earlier, the normal motion in canals is turbulent and head losses are proportional to the square of velocity and therefore respond to the general formula $P = k \cdot v^2$. What is needed is to know what value should be adopted for k in each case.

The most frequent points where load losses occur are curves.

Boussinesq's formula is:

$$J = \frac{v^2}{(R \cdot c^2)}(1 + \frac{3}{4}\frac{\sqrt{b}}{\sqrt{r}})$$

and is applicable for calculation, where

J is the total hydraulic slope by unit length in curve
v is the speed of water, in m/sec
R is the hydraulic radius of the canal section, in m
c is the Chezy coefficient
b is the maximum width of the canal section
r is the radius of the curve.

It should be pointed out that the term $v^2/(R \cdot c^2)$ is the gradient (according to Chezy's formula), which therefore is known in the canal we are designing. The sum in the second parenthesis indicates the hydraulic gradient excess caused by unit of length in the curve and which must be added to the normal slope.

This means that the loss of total load is figured by multiplying the value given by the formula by the length of the curve.

The loss due to the curve effect increases with the square root of the quotient b/r. This makes it advisable to adopt large radii not less than five times the width. The square root will then be less than 0.44 and the 3/4 factor reduces the loss from the loss of curve load to 0.33 of the normal slope.

As the length of each curve, measured in kilometres, is very small (we mentioned earlier that the normal slope losses per kilometre in canals are very slight), the losses in absolute value are usually very small on each curve, except in cases of high-velocity sections, which exist in canals for hydroelectric use, but are infrequent in irrigation canals.

Therefore, it is usually acceptable in many instances not to take into account the localized losses on curves and include them in an average value throughout the entire canal, based on reasonably increasing the roughness coefficient of the canal. Nevertheless, there will be instances when the individual calculation will be advisable, because of either high speeds or very small radii.

Another source of localized losses are the syphons. Since these will be the subject of a special study to be made further on, we will not discuss their calculation until that time.

Section changes are frequent in canals and often go hand in hand with changes in speed. The load losses in these transitions are expressed by the general formula:

$$P = k \cdot (v_1^2 - v_2^2)$$

where v_1 and v_2 are the two rates of speed. The value of k depends on the type of transition and its length.

As a general rule, it can be said that the length of the transition should be about five times the difference in maximum widths of the canal, which is the same as saying that the total angle of transition is some 22 or 25 degrees. Under these conditions the losses are slight.

When the canal narrows and the length of transition meets the prior condition, the head loss is almost none.

To the contrary, a less favourable case is the widening of the canal. Even in cases of smooth widening, values of the order of 0.35–0.50 must be assigned to the coefficient k.

Lastly, the gates are a source of other localized losses in canals. Naturally, one of the purposes of the gates is to more or less intercept the passage of water when they are closed or semi-closed, which is equivalent to the insertion of a strong head loss. But at this moment we are referring instead to the losses caused even when the gates are totally open, due to the section changes made in the canal for its implementation. In each instance, the calculation should be made bearing in mind the section variations and how brusque the transition is, in accordance with what was said earlier and knowledge drawn from books on hydraulic theory, which will have to be consulted in some specific instances.

In the case of partially closed gates, the water upstream stops flowing and the water level differs depending on the conditions downstream.

The formula that gives us the flow circulating through a submerged hole is:

$$Q = c \cdot A \cdot \sqrt{2 \cdot g \cdot h_1}$$

in which

Q is flow in m^3/sec
A is area of the hole in m^2
g is acceleration due to gravity $= 9.8 \, m/sec^2$
h_1 is the depth of the axis of the hole under the water-free level above, if the exit is free or else the difference between upstream and downstream water level if the exit of the jet is submerged.
c is a contraction coefficient which, in the absence of other criteria, may be about 0.6 or 0.7 (89).

Gates may either be submerged (like outlets) or not.

When the opening they cause is a submerged hole, the prior formula is applicable. Calling y and $y - y_2$ the upstream and downstream water depths, respectively, measured with the same datum, the applicable formula is:

$$Q = c \cdot A \cdot \sqrt{2 \cdot g \cdot y_2}$$

An important case is that in which a structure inserted in the canal causes a difference in level between upstream and downstream waters, leading to a flow of water through the obstacle that may be broken down into two parts: one that must overcome the pressure of the downstream level and another that overflows freely. (For example, the pillars of a bridge placed in the water, or a submerged dam above which the water passes.) (Figure 1.2).

In this case, a close enough estimate may be made of the flow running at counterpressure by means of the formula just cited of submerged orifice

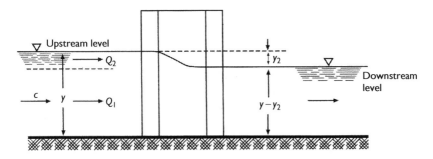

Figure 1.2 Mixed flow: spill and counter-pressure.

and the free-falling flow of the upper part, by means of the Weiz formula, which is:

$$Q = cL \cdot (y_2)^{3/2}$$

in which c is a coefficient of about 2.00.

> L is the useful width of the canal at that point
> y and $y - y_2$ are the depths upstream and downstream of the obstacle respectively.

At times grids must be placed at the entrance of the canal, in front of large syphons, etc. They are usually inclined to facilitate cleaning.

The formula for the head loss is:

$$h = b \sin \alpha \left(\frac{d}{a}\right)^{4/3} \left(\frac{v^2}{2g}\right)$$

in which

> h is head loss
> α is the angle of the grid with the horizontal
> d is the diameter of the bar, in cm
> a is the free light between bars in cm
> v is the velocity in the vertical section prior to the grate
> g is the acceleration due to gravity
> b is the coefficient depending on the shape of the braces, as follows:

> > $b = 1.79$ if round in shape
> > $b = 2.42$ if rectangular in shape
> > $b = 1.67$ if rectangular with rounded ends
> > $b = 0.76$ for airplane-wing shape.

1.5   Variation in traction throughout the section

The speed of water is not uniform in the entire cross section of the canal. Rubbing against the walls makes the water closest to the latter move more slowly. The braking effect is gradually transmitted to the water farthest away until the highest-speed area is reached.

The speed deduced from the aforementioned formulas of Chezy, Manning, Bazin, etc. is the average speed.

If the points of the same speed in the cross section of the canal are joined into a line t, the resulting curves are called ISOTACHS (in Greek this means they have the same speed) and may be as shown in Figure 1.3.

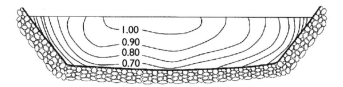

Figure 1.3 Isotachs curves.

If we imagine we are tracing the family of curves orthogonal to the isotachs, we will have the canal divided into a series of primary canals bordered by two contiguous orthogonal lines, the bottom and the surface of the canal.

Each of these primary canals is characterized by the fact that the water it carries rubs only against the floor of the canal, since laterally it borders on another canal and at each point of contact, the water particles of each one travel at the same rate of speed (because they are in the same isotach), without either one slowing down the other.

For each one of them the traction formula $t = 1000 \cdot R \cdot s$ is still valid. But if the canal is regular in shape, the sections of each primary canal have a width that varies little and is relatively similar from one to another; therefore, the size of their surface is approximately proportional to the depth of each canal. The wet perimeter, which, as indicated earlier, is limited to the part in contact with the floor, varies little from one primary canal to another. As a result, the hydraulic radius of each one is almost proportional to the respective depths.

The tractive force for each one is equal to $t = 1000 \cdot y \cdot s/\cos \alpha$, where y is the depth of each one and α the angle of the slope.

The result is, speaking in fairly approximate terms and as proven experimentally, that the traction in a trapezoidal canal increases on the slopes in a lineal manner from zero in the upper part to a maximum value at a point close to the bottom, where it is almost constant (Figure 1.4, according to studies conducted by the Bureau of Reclamation).

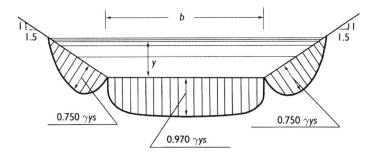

Figure 1.4 Tractive force along the perimeter of the canal.

In the areas next to points where the bottom joins the slopes, there is a notable decrease in tractive force. The reason is that at those points the water particles are exposed to a great deal of rubbing (at the slope and the bottom), and therefore they move at very little speed (which is reflected in the shape of the isotach curves). The decrease in speed, as we already know, causes a decrease in tractive force (shear stress).

1.6 Stable canal forms

Tractive force (shear stress) is different at each point of the cross section of the canal and is known, as we have just seen. But the stability of the particles on the canal's perimeter depends both on erosion caused by tractive forces and on the stability caused by the incline of the area where the particles are located. Obviously, the particles on the inclines, which may roll downhill, are in a worse situation than those on the bottom, which are on a more stable, horizontal surface.

From the start one guesses that the shape of the cross section of an unlined canal, in which all points have the same stability (stable canal form), is a curved surface that near the bottom (where the depth and tractive force are greater) has a very mild transversal slope, next to the horizontal line in order to give it greater stability. To the contrary, in the lateral areas next to the bank of the canal, the water depth and the tractive force are very small, and therefore the slopes are more inclined.

There have been many authors who have studied the stable canal form. For example, Henderson (49) or Leliavsky (60) who concludes that its equation is as follows

$$y = k \cdot \cos(x/c)$$

where x and y are the coordinates of any point in its perimeter with respect to rectangular axes originating at the centre of the free surface of the canal. The values k and c are constants that depend on the maximum flow and the angle of rubbing against the terrain. We will go into this more deeply in Sections 11.4 and 11.5.

The shape of the stable cross section is therefore that of the function cosines. The parabolic forms and, even more so, those of incomplete circular arc are quite similar, and, therefore, it can be said, in approximate terms, that the latter are very stable, making this one of their advantages.

In strictly theoretical terms, the ideal would be to have unlined canals (which are those most affected by tractive forces) with stable forms. This would be like having houses with pillars and beams designed with the same safety coefficient instead of, for example, having beams with too much resistance and others with a strict degree of resistance. Still, there are other reasons for not always adopting stable forms, which we will study in due course.

1.7 Subcritical and supercritical motion

A disturbance in the water of a pool (the waves caused by throwing a stone into the water, for example) is transmitted by the so-called "celerity" (wave velocity), which in theoretical hydraulics is equal to $\sqrt{g \cdot y}$, where g is the acceleration due to gravity and y is the water depth.

If, instead of a pool, the question is a channel of moving water, any disturbance in the water at any point in the channel is transmitted downstream at a speed equal to the sum of that of the canal plus the wave speed, and upstream at a speed equal to the difference between the wave speed and that of the canal itself. But if the speed of the canal is greater than the wave speed the disturbance cannot be transmitted upstream, since the greater speed of the canal drags the disturbance downstream.

This property is well known but is essential in order to be able to analyse the variation in the surface of a canal. If the rate of speed is slow (i.e. less than the wave speed), the temporary disturbances caused by gates, derivative intakes, change in the operation of turbines, etc. influence in a downstream to upstream direction, and, therefore, the study of the variation in the surface of the canal should be made beginning downstream.

If it is a matter of a section of canal in supercritical motion (speed higher than celerity), the upstream part suffers no influence at all from temporary variations in the downstream area; quite the contrary, it is the upstream part that governs downstream movement and, therefore, the study of surface variation should be made in the direction of the water movement.

In a canal with exactly the critical rate of speed (equal to celerity), the waves of the possible disturbances would travel in the direction of the water, but the wave part that tends to move upstream would have a stationary front and would remain immobile, since its rate of speed would be equal to and contrary to that of the canal.

The critical rate of speed is $\sqrt{g \cdot y}$, in which y is the critical depth. The critical rate of speed, if the cross section of the canal is not rectangular, is the same as the former but for a fictitious average depth which is A/T, in which A is the area of the section and T the greatest width. In other words, it is $\sqrt{g \cdot \frac{A}{T}}$.

An interesting concept is that of the critical depth: that for which the resulting speed in the canal, for a predetermined rate of flow, is the critical speed. Assuming a predetermined water depth, the critical slope is that which originates the critical speed.

If, in the design of the canal, the slope goes from values that cause a low rate of speed (subcritical) to others that cause a high rate of speed (supercritical), the change always occurs at a point where the existing rate of speed is the critical one. If possible, the canal should be designed in such a way that the critical speed occurs at an appropriate point, which is called "control section". See Section 15.4.

If, to the contrary, the water goes from a supercritical to a subcritical rate of speed, this almost always occurs by means of the formation of a hydraulic jump (see Section 15.4), and its study is found in virtually all the literature on hydraulics (15), (16), (49), (98).

The functioning of hydraulic jump is based on two fundamental equations: the preservation of the water mass (what comes in goes out) and the preservation of the amount of movement, since there are no exterior forces to the jump, just the interior forces that act between the two parts with a distinct depth.

If we call y_1 and y_2 the water depths in metres before and after the hydraulic jump, q the rate of flow per metre of width in m³/sec. (we are assuming a rectangular canal), the formula that relates the two depths is

$$y_2 = \frac{y_1}{2}\left[-1 + \sqrt{1 + \frac{8q^2}{(g/y_1^3)}}\right]$$

The water depth y_2 is called a conjugate of y_1 and is necessary in order for the hydraulic jump to occur. If it is smaller, the hydraulic jump does not occur and if it is greater, the hydraulic jump is drown and it is called "submerged jump". For canals with trapezoidal cross section, the formulas are more complicated and not customarily used in practice since an approximate calculation with equivalent rectangular sections is sufficient. At any rate, they can be seen in the literature on theoretical hydraulics (35), (98).

The hydraulic jump may be successfully formed at a predetermined point in the canal (if there is supercritical speed and the water depth downstream is less than the estimated conjugate) by lowering the canal floor to make the water depth somewhat greater than the required conjugate depth (it is always a good idea to adopt a safety factor).

The length of the hydraulic jump depends on the Froude number, and is always from 3.2 to 5.5 times the conjugate water depth downstream. An interesting study on details of the hydraulic jump can be seen in an excellent work by the deceased engineer Luis Torrent (93). On the other hand, in canals, which are linear works, there are almost always enough length available. At any rate, if one wants to know more precisely the length of the hydraulic jump, there are publications that give suitable charts (29).

The hydraulic jump is generally the best method for lessening the kinetic energy of a current of water and is, therefore, the appropriate system for connecting the end of a reach of steep slope with a mild-slope reach, thus avoiding possible erosion. Nevertheless, the percent decrease in energy depends on the nature of the hydraulic jump. In the case of canals, whose rapid sections usually have Froude numbers of from 1 to 3, the energy loss is no more than 30%, less than those obtainable in dam spillways, where the Froude numbers are quite a bit higher. Let us recall that the Froude number is $F = V_1/(g \cdot y_1)^{0.5}$, in which V_1 is the velocity and y_1 the depth of water.

It may be said a priori that if the hydraullic jump is covered with water, the rate of energy loss will be less. In the publications mentioned earlier, there are methods for figuring the possible reduction in kinetic energy exactly.

1.8 Stable varied flow

Using the formulas cited in Section 1.3, regarding levels and flows and localized losses of charge, a canal can be designed with a predetermined, uniform flow (with the surface parallel to the bottom) and constant over time. But for other rates of flow, even though constant over time, the surface in that canal would not be parallel to the bottom, since the head losses in the syphons, bridges and other singular points will be different and will no longer be in accord with the levels given by the Manning or similar formulas. Therefore, backwaters will be formed. For these cases, varied regime formulas must be utilized. Such formulas consider a regime that is constant over time but varies along the length of the canal, with the surface not parallel to the floor.

These formulas are based on the fact that the head loss due to varied movement of the water is equal to the algebraic sum of that caused by hydraulic friction plus that required by the variation in kinetic energy.

Calling Δx a short distance along the bottom, Δy the corresponding variation in level, s the slope, c the Chezy coefficient (deducible from the table shown in 1.3 for the Manning and Bazin coefficients), Q the flow at each point, P and A the perimeter and area of the cross section, as average values of the section included in the Δx interval; A_1 and A_2 the value of the area at the ends of the canal section and α a coefficient of losses of kinetic energy, the formula for gradually varied movement is:

$$\Delta x = \frac{\Delta y - \alpha \cdot \dfrac{Q^2}{2.g} \cdot \left(\dfrac{1}{A_1^2} - \dfrac{1}{A_2^2} \right)}{s - \left(\dfrac{Q}{c} \right)^2 \cdot \dfrac{P}{A^3}}$$

bearing in mind that in each calculation of a new section, one must figure P, A, c, A_1, and A_2 of the new end of the canal stretch being considered.

Using this formula, the levels of this curve, called "backwater", can be calculated. Starting from a point where the water depth is known (e.g. upstream of a half-closed gate which causes a backwater), other successive slightly lower water depths are considered and each one of the values that is part of the formula is calculated. The formula gives us the distance Δx to the point which has the considered water depth.

Full information on this topic can be seen in numerous books on hydraulic theory (49), (80), (98).

At present it is easy to make or find computer programs to perform this calculation. Programs that can do the general study of a complete canal, including all kinds of inserted singular works by the Bureau of Reclamation, should be mentioned (21).

The water surface types that may occur in a canal in the gradually varied movement are the subject of detailed study in many books (15), (16), (49), (89), (98). Curves corresponding to gentle slopes (subcritical) are usually called "M-type" curves and "S-type" curves are those corresponding to supercritical (steep) slopes, see Figure 1.5.

The M_1 curve occurs when, in a slow-moving section, the water rises over a gate or for any other reason there is a narrowing or localized head loss downstream.

The M_2 curve occurs when water flowing at a slow rate speeds up because it is approaching a section with a greater slope or drop or a syphon calculated for a greater flow than that existing at that moment.

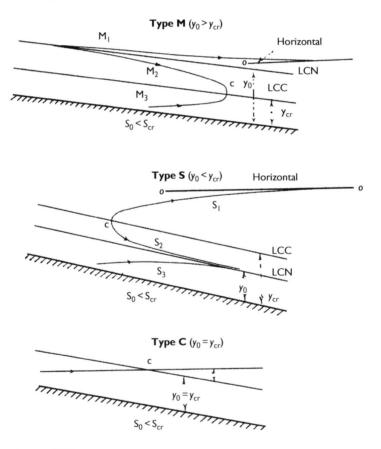

Figure 1.5 Water surface types.

The M_3 curve occurs when water flows out from under a gate, with a mild slope, at a high rate of speed and gradually slows down until it forms a hydraulic jump.

Canals with a supercritical slope are less frequent and therefore the S curves as well. The S_1 curve occurs as a high rate of speed slows down because of a gate inserted in the canal.

The S_2 curve occurs when slow-moving water enters an area with a supercritical slope and speeds up until it reaches critical speed.

The S_3 curve occurs when water moving at a fast rate of speed enters, e.g. a section of canal with a lesser slope (though also supercritical).

It should be pointed out that these types of surface generally occur only in a stretch of those indicated in the figure, without its being necessary to include their beginning or end.

1.9   Unsteady flow

In a canal the rates of flow and levels forcibly vary over time, as required by the supply conditions. The flows feeding into the canal vary, as do the exit flows, the superelevation of the gates, etc.

In such a situation the calculation of a canal and the variations in the surface over time cannot be done as indicated thus far. One must take into account the variations that occur over time, at times quite rapidly (closing or opening of gates).

The equations that govern the water movement are the well-known Saint-Venant equations. There are two of them: one requires that the rate of flow be preserved (the water is neither created nor destroyed), and the other the preservation of the amount of movement, taking into account that this value is affected not only by the change in speed but also by the change in the water mass found in the section being studied.

The cited equations, in partial derivatives of space and time, are as follows:

$$\frac{\delta y}{\delta t} + \delta \cdot \frac{v \cdot y}{\delta x} = 0$$

$$\frac{\delta v}{\delta t} + v \cdot \frac{\delta v}{\delta x} + g \cdot \frac{\delta y}{\delta x} = g \cdot (s - sf)$$

in which

  y is the depth of water
  t is the time
  v is the speed
  x is the distance
  s is the slope of the canal floor
  sf is the dynamic slope

i.e. that which complies with the Manning formula if the real speed and hydraulic data are introduced therein.

The system of both equations should be complied with at all points and at all times, which means that the non-uniform movement in a canal over time can only be calculated with a computer program. More information on this point can be found in Chapter 20.

Chapter 2

Water loss in canals

2.1 Worldwide importance

According to an interesting study carried out by the International Commission for Irrigation and Drainage (56), around a third of the water that uses irrigation canals all over the world is lost during transportation. Another third of the water is lost on the plot of land or the farm when it is used for irrigation. Therefore, only one-third may be usefully applied.

These horrifying figures refer to worldwide average values, since there are cases, such as those quoted by Kennedy in 1981, where canals lose 45% of the water they are transporting.

There are the individual cases in the USA where losses during water transporting in canals reach 60% and on average they reach 23%.

In many countries throughout the world water is scarce, therefore it must be rationally used. The cases of North Eastern Brazil, South Eastern Spain, areas in the Middle East, etc. where the lack of water restricts production are only examples known by the public in general, but they form part of a much more widespread problem.

Canals used for town water supply or for hydroelectrical use also lose a significant amount, although we do not have any statistics for them.

When water is lost in canals we must ask ourselves what benefits would have been obtained with the construction from the beginning of canals that were less permeable or the improvement of the watertightness of existing ones and therefore the profitability of this investment.

In the case of irrigation, for example, it is clear that saving of these water losses would mean irrigation in new areas, with significant increase in production. But similar considerations may be made for energy production canals, for town water supply canals, etc.

In the cases of using purified waste water, desalinization of sea water, expensive water transfer work from other basins, the interest in decreasing the losses to the maximum economically possible is obvious, since a significant amount of water could be saved and could be transferred to other uses.

Moreover, even in those countries where there is no lack of water, its catchment, regulation and even more its elevation are expensive operations, therefore preventing useless losses during the transport is an economically interesting operation.

You only have to imagine a canal that has to transport a certain amount of water to a certain place. If, during this operation, it suffers from significant losses, the water flow that we have to introduce through its main intake must be greater, and equal to the sum of the required flow and the predicted leakage amount. This without doubt makes not only the energy to raise water from a well (if this is the supply system) but also the construction of a possible regulating reservoir (if surface water is taken) and even the work on the canal itself more expensive, which must have a greater transport capacity, since as well as the final water requirements, it must transport the leakage itself that will be lost en route.

Any of these considerations have justified the decision to line canals to decrease their losses. In Spanish legislation there have been many economic aid rules for irrigators, wanting to line their tertiary irrigation canals. The Decree of the 15 December 1939 is a simple example. The water saving that it represented and the profitability its use on other applications meant it easily compensated the State for the economic expenditure carried out. Another important example is that of the Water Supply Consortium of Tarragona, which used a flow of almost $5\,\mathrm{m^3/s}$ saved thanks to the lining of the canals of the Bajo Ebro.

When reaching this point an important question arises: given that the civil work that we can build, repair and handle cannot be perfect and will always have some losses, what maximum value can we aspire to and what technical possibilities have we to obtain this?

Frequently it is considered that a canal that has losses due to leakage between 25 and $50\,\mathrm{l/m^2}$ in 24 hours adequately fulfils its undertaking. This value is not clearly defined, but it represents an order of magnitude that is generally internationally accepted (56).

It is obvious that many canals, because they are located in highly impermeable land or because they are covered with linings that are very well conceived and very well made, can have lower leakage rates (at times almost none), but it is no less true that in many cases, if greater precautions are not taken, the losses are much greater, due to the high permeability of the land, due to performance faults or simply due to cracking of the linings. This demands a very delicate care by the project engineer to study the solution to be adopted and to carefully specify the conditions that must be complied with during the work. Only in this way may the leaks be kept within the abovementioned minimums.

If, to bring our ideas together, we consider a real canal that transports $60\,\mathrm{m^3/s}$, with a speed of $1.5\,\mathrm{m/s}$, a wet perimeter of $20\,\mathrm{m^2}$ per lineal metre and total length of $100\,\mathrm{km}$, the leakage surface area (wet perimeter) in the case of the dimensions of the cross section being kept throughout the layout would be around two million square metres. If the canal were telescopic, the wet surface area would be half of this, i.e. one million square metres. With the abovementioned leakage values, the losses would be from 25 000 to 50 000 m³ per day.

The volume of water transported by this canal per day will be 60 m³/s × 86 400 s/day, i.e. 5 140 000 m³; therefore the relative loss is around 0.5–1%.

If the water speed were higher (as normally happens with canals in hydro-electric plants), the loss percentage would be even lower.

This percentage might seem small, but other causes that increase it considerably must be taken into account.

A canal such as this one that supplies an irrigation area supplies a network of irrigation canals and derived canals the length of which is around ten times greater (in the specific case that we are talking about, the length of the canal represents 2 m/ha and the secondary channels around 20 m/ha). Though the latter are much smaller (a tenth in size of the canal), their losses due to leakage are approximately equivalent to the losses of the main canal and the total losses are doubled.

There is another important cause for water loss in canals, unfortunately about which we can do nothing. Here we are referring to losses due to evaporation.

In the real canal we are considering, the surface width is approximately 15 m; therefore the evaporation surface area over the 100 km of the canal will be 1 500 000 m² or 150 ha, which would be reduced by half if the canal were to be telescopic. With a fictitious evaporation of 1 l/s per ha (which can be real on many occasions, but which in general errs on the side of safety), the loss would be 150 l/s, equivalent to 12 960 m³/day, or half of this if the canal were telescopic, i.e. a value around a quarter or eighth part of the values to which the losses would be reduced if we were to limit the leakages to 25–50 l/m² in 24 hours.

We have no weapons to fight against evaporation. But we have to be aware that the maximum desired leakage of 25–50 l/m² in 24 hours is equivalent to a loss percentage of around 1–2%, to which another 0.5% must be added due to evaporation.

There is a third cause that produces water losses in canals that is worth mentioning. We are referring to involuntary spills due to problems of bad regulation in the handling of the canal. It is a subject we will talk about in this book and that requires suitable facilities to be foreseen in the canal and careful planning in its handling. With the correct precautions the losses due to this reason, together with the leakage, make a total of 5% of the water flow, which is an admissible value, particularly if we compare it to the 33% that we have given as the current average in the world.

In this book we will try to fight against losses due to leakage and due to unsuitable spills.

However, since we are aware that we will not be able to design and construct canals in general that have no losses, we must oversize their capacity to carry not only desired flow, but also the losses. Traditionally, with a safety margin and whenever great care in the design and the building of the canal were taken, this concept means the calculated flow is increased by 10%.

2.2 Causes that affect water losses due to leakage

The maximum losses due to leakage in 24 hours to which we aspire are equivalent to a water height of 2.5–5 cm over the entire wet inner surface area of the canal, a value that for many will not seem negligible.

The fact that different causes can affect the leakage value in a canal must be taken into account.

The most important one is the permeability of the land (more, or less, clayey) and together with this, the quality of the lining (if there is any) and its conditions.

The cross-sectional shape of the canal is also important. A big depth makes the hydrostatic pressure on the bottom increase the losses, assuming naturally a water table that is below the bottom of the canal. As we want to limit the losses due to leakage to $25–50 \, l/m^2$ in 24 hours, it is obvious that the total loss will be greater in larger canals than in those with a limited size, not only due to their greater leakage surface area, but also due to the greater leakage pressure that a big depth implies.

The amount of sediment in suspension or carried in the water is highly significant. If there is any loam, as the water leaks through the land it drags fine particles that little by little sediment between the holes in the land, silting it up and decreasing its permeability. On the other hand, very clear water, when it leaks, drags finer particles from the soil, making the permeability increase. From this point of view, surface water from a not-so-high-mountain river is much better regarding leakage than water from a well, where the water is pumped after a significant leakage into the land.

The frequency of the canal's periods of use also affects the leakages. It must not be forgotten that the soil swells when it gets damp and on the other hand, when it dries off, it retracts. In the retraction periods, cracks appear, which although frequently seem tiny aid leakage. From this point of view it is advisable to keep the canal full whenever possible, to avoid changes in the dampness of the surrounding land. We have known engineers who build canals from their main inlet to the downstream area, supplying water from upstream to the new stretches and keeping them with still water, in an attempt to keep the same humidity from the very birth of the canal.

The age of the canal is a highly important piece of information. As with all beings and things, a canal will deteriorate as time goes by.

Weather conditions also affect canals, not only because the weather might be better or worse and therefore, more difficult to bear without deterioration, but because during dry periods the land evaporates water more easily and leakage increases to replace this loss of humidity.

The roots of vegetation around the canal dislocate the soil structure, creating leakage routes. For aesthetic reasons, tree plantations are made alongside canals and particularly along service roads, since they are easy to maintain, with irrigation from the canal itself. These should not be rejected, but they should be located at a reasonable distance from the canal.

2.3 The way to stop leakage in a canal: Other reasons for putting in linings

The maximum values for losses due to leakage must be common to all canals. If the soil itself already ensures lower losses, a canal may be built without lining, if the conditions discussed in Chapter 11 for unlined canals are also fulfilled. But if it is foreseen that the leakages are going to be greater than those admitted, the most common solution is to line the canal.

The lining is nothing more than a layer of long-lasting material that is placed over the excavated and profiled surface area of the canal. Taking into account that its purpose is to decrease the canal's leakages, it is obvious that its basic characteristic must be its waterproofness. This solution means that the canal's excavation slopes are perfectly stable, but the thin lining must not be expected to support them.

The decision as to whether a canal should be lined or not, to improve its waterproofness, must be taken before its construction, after a study of the characteristics of the land. If a well or hole is built, with specific measurements, next to the layout of the future canal and with the same depth, the future leakage may be predicted. It only has to be filled with water, and after 24 hours cube the primitive volume and the reduction in its level to calculate the leaked volume. Dividing this by the wet surface area of the well, we will have a value that, according to whether it is lesser or greater than 25 or $50 \, l/m^2$ in 24 hours, will indicate whether the soil is sufficiently waterproof or not.

If, on the other hand, the canal has already been built, there are other systems to evaluate the importance of the losses, although they are not always as accurate as could be wished for. One of them consists of transforming the canal into a series of successive reservoirs formed by some small provisional dams that are built within the canal and that are filled with water. After a certain time (for example, 24 hours) the drops in the levels are measured and the leakage per square metre may be calculated. There is the possible disadvantage that the dams (or some of them) may not be waterproof and let water through, meaning that the results may be hidden, at least in the localized results. Another disadvantage that cannot be forgotten is that to carry out this study, the canal must be taken out of working order. Obviously, the canal's leakage surface area must also be calculated, since as the levels of the reservoirs are horizontal, it is a different surface area to the one normally wet when the canal is in use.

Another system, more frequently used, consists of measuring the water speeds in the canal during normal service, using a current meter. As the speeds at different points of the cross section of the canal are not equal, several measurements must be made at each section and the results must be integrated. This system is not valid when there are intercalated derivations, unless these are endowed with reliable measurement data. The differences in the water flow between different stretches give us the existing leakages.

Specific leakages may be measured in a full canal using a piece of apparatus similar to that shown in Figure 2.1 (83).

This ingenious device measures the leakage through a canal surface lined by a plate. The water leaked to the soil does not come from the canal; it

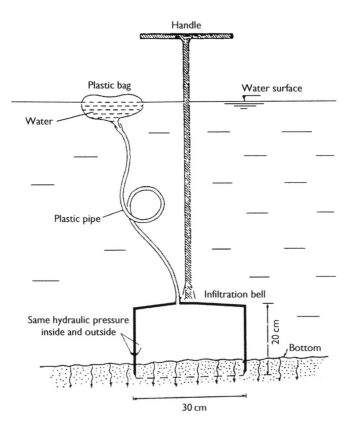

Figure 2.1 Device for measuring specific leakages (according to Florentino Santos).

comes from a rubber bag the contents of which have been previously measured. This device measures the leakage at a specific point. It is a drawback because frequently the leakages in lined canals are highly irregular, being very high in the cracks and breakages and non-existent in the places where the lining is in good condition. It could be useful in unlined canals (with homogeneous characteristics), etc.

The lining may be made, as we will see under Section 2.5, from very different materials. Some of the characteristics may be very interesting to resolve certain problems in canals, therefore at times canals are lined for reasons other than waterproofing.

The most important cause (after waterproofing) for lining a canal is to achieve a greater resistance to water erosion, an important problem in many canals in soil, as we will see in Chapter 11.

Another possible justification is the fact that a lined canal has a better roughness coefficient and, therefore, a greater transport capacity. We must remember the so-called "hydraulic paradox" according to which on lining canal, occupying part of its useful section with the lining, its flow, however, is increased because the percentage increase in velocity is greater than the percentage decrease in the cross section.

Vegetation growing in a canal usually presents important problems, which enforce frequent, problematic cleaning operations or "trims". A lining makes it difficult for vegetation to grow and encourages cleanliness.

2.4 Canals that must not be lined: Drainage canals

Section 2.3 shows the cases when it is advisable to line canals and others when it is not necessary or it is not justified. However, there is one type of canal where lining is particularly advised against. We are referring to drainage canals, the purpose of which is not to move water from a supply point to a consumption centre (irrigation plot, water supply for a town, hydroelectric station that takes advantage of available waterfall, etc.), but rather for the canals to collect and take the water drained from irrigated or soaked land to an evacuation place.

If these canals were to be lined, they would have great exposures to under-pressure problems, due to the water table that soaks the land, the possible flows being so large that the drainage techniques for the lining that we will show in Chapter 9 would not be applicable.

These canals must be unlined, Figure 2.2, to be able to freely admit the water that leaks from the land along its slopes, which may lead to other drainage canals that are tributaries of them. In this way, its drainage action is complemented.

The old Irrigation Canal Standards from the 1940s, drafted by the then Ministry of Public Works of Spain, recommended trapezoidal cross sections for canals, whose horizontal bottom was lined with concrete on site. This

Figure 2.2 Drainage canal.

cross section had the advantage of aiding the cleaning of sediments on the bottom and not preventing the leakage of water through the slopes.

It is highly probable that when projecting drainage canals we find problems of excessive speeds. As we cannot fight these with linings, a series of drops (described under Section 17.3) must be introduced into the canal, which allow the initial gradient to be replaced by a series of stretches of canal with smoother gradients and with less danger of erosion.

As they are canals made in soil, with erosion danger, they are often endowed with concrete transverse sills, embedded into the ground (as may be seen in the Figure 2.2), a technique that is also frequently used to stabilize river beds, since they establish points on the bottom that have a contour height fixed by the transverse beam (Figure 2.2).

Drainage canals may have greater erosion where tributaries join them. A short stretch of the largest canal should be lined to prevent erosion due to the drop off.

2.5   Types of lining

The most commonly used lining in the world is bulk concrete made on site. The currently used thickness is between 8 and 15 cm. Its waterproofness is magnificent, whenever it does not crack due to thermal forces or earth movements. Lesser thicknesses than those indicated (with cement mortar) are not recommended since they crack and deteriorate easily (Chapter 4).

Sometimes linings of concrete reinforced with iron bars are used to combat cracking (33) (Figure 2.3). The main objection is its greater cost in relation to bulk concrete, not only due to the price of the iron, but also due to the fact that it is more expensive to lay. It is usually only used in special cases, which we will indicate in this book when necessary.

It should be mentioned that, as the lining is thin, it is difficult to guarantee the distances required from the iron bars to the concrete surface with sufficient accuracy. If building errors occur on this point, the lining could resist the possible bending moments less than could be expected.

The report by Chester W. Jones from the Bureau of Reclamation (35) is very interesting, where he explains that this organism decided not to reinforce the concrete coverings with iron bars in 1948, except in special case, obtaining a saving of 10–15%. The same is stated in (25) and (32).

Reinforced concrete is more suitable for thick linings (which should really be called "retaining walls"), which must also resist thrust from the outer land or the inner water. In these cases it can be a useful material.

Concrete may also be used to make prefabricated slabs to line the canal. It is a lining with special characteristics that we will analyse in depth in Chapter 6.

Rubblework linings (thick stones jointed with cement mortar) are infrequently used; however, they do have a field of application, as we will see in Chapter 6.3.

Brick linings are indicated in countries with abundant clay soil and cheap labour that allow low-cost manufacturing.

Linings are made with bituminous agglomerate (also called "asphalt mix"). They are similar to road pavement and they have the advantage over bulk concrete linings in the fact that they are more flexible (relatively), which allows them to bear the dilations and retractions due to changes in temperature and the flexions due to land settling better.

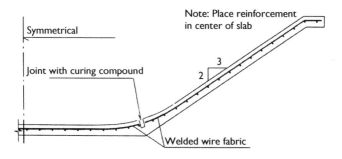

Figure 2.3 Reinforced concrete lining.

Even more flexible are the membranes, made of different materials. They are studied in Chapter 8.

Improved soil linings are highly interesting, formed by mixing natural soil with other soil with better characteristics that is transported from another place and then mixing it for waterproofing determinants and erosion resistance. It is really a canal in soil, but with higher quality than natural soil (Chapter 11).

Other waterproofing systems or land protections may be obtained using different procedures (soil-cement, bentonite, chemical products, etc.; Section 11.12).

The choice of lining depends on many reasons, amongst which the most important are the availability of materials, the cost of labour and of the auxiliary machinery, the level of waterproofing required with the lining, the maintenance expenses and the efforts to which it is going to be subjected, such as the water velocity or soil thrusts, changes in temperature, growth of vegetation, etc.

In the following chapters we will study the most important characteristics of the aforementioned linings, in order for engineers, in each case, with full knowledge of the facts, to choose the most suitable one, repeating that, in any case, the efforts that are going to act on it are always very hard.

There are more canals in the world that are excavated in soil without linings. Lower cost has a great influence on this fact. They can give good results if they are well designed and built, a fact that does not often happen.

Most lined canals are those lined in bulk concrete.

Although the statistics in the USA are not representative of the world reality, due to their different technology and particularly the different economy, the percentage of the canal length for each type that was built in this country by the Bureau of Reclamation between 1963 and 1986 (30) could be interesting:

Reinforced and non-reinforced concrete	58%
Compact soil	27%
Membranes and mastic linings	7%

It is worth mentioning that the canals that appear as compact soil are not truly canals only excavated in soil, rather they refer to canals with improved soil linings.

2.6 Resistant side walls

Sometimes canals are made with a bottom and sides that, as well as preventing water losses, are resistant elements. Often they are called "retaining walls". The basic difference with the linings is the fact that in these the resistant element to the hydraulic pressure is the soil, the lining being merely a protective or waterproofing factor. The retaining walls, on the other hand, are used to resist the interior hydraulic thrust on the canal and often also the thrusts of the soil itself. They are therefore structural elements the resistance of which must be calculated as opposed to all the possible combinations of exterior efforts.

They may be made in bulk (concrete or rubble work, infrequently in brick) or reinforced concrete.

If they are made of rubble work, brick or bulk concrete, their stability relies on the weight itself, as if they were small gravity dams and the same that they must also resist the possible uplift pressure due, in this case, to the canal water that may be leaked under its support base (Figure 2.4).

There are several situations when retaining walls are used, which are listed below:

a   When the water level must be above the land level. In this situation two different solutions may be adopted: With resistant walls (Figure 2.4) or with a canal on an embankment (Figure 3.2 (third diagram)). In this situation a comparative study must be carried out, to choose among the technical solutions with a view for the economic studies.

b   When the land has a highly unreliable resistance, and slopes that are too inclined would have to be adopted.

Figure 2.4 Cross section of a retaining wall.

c   When the natural land's cross-sectional slope is too great or unstable, so that it would not stand the positioning of a canal with a smoother slope as may be seen in Figure 2.5, which is a photograph of the Canal of Las Dehesas (Extremadura, Spain).

When the resistant walls are made in bulk, they usually have the cross section similar to that of a gravity dam. We know that these are stable when they have, for example in triangular profile, the front vertical slope and the back slope of around 0.75. We can give the resistant walls a similar cross section, but in the calculation of these, due to their size, we cannot forget their greater width of the upper part, which must have a minimum value that is enough to be able to introduce the fresh concrete within the formwork of the resistant wall. The volume will be a little greater than if the resistant wall were to be triangular, and so will be its resistance, but we can reduce them by decreasing the back slope a little, making some simple calculations that obviously must take a possible uplift pressure into account.

They must also resist the sliding effects produced by the inner water. As the resisting walls are usually supported on the previously made bottom, if concrete is used as a material, a sufficient link between the bottom and the box cannot be expected, due to the fact that fresh concrete never welds well with concrete that is already dry. Therefore sometimes some steps are made on the bottom and retaining wall (Figure 4.13).

Figure 2.5 Canal on land with a strong cross slope (*Canal of Las Dehesas* (Extremadura, Spain)).

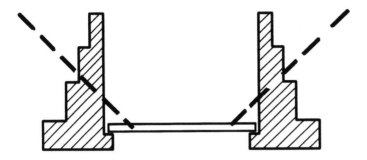

Figure 2.6 Stepped resistant wall.

If the retaining wall must resist the soil thrusts, we can give it the same previous cross section or a stepped one like the one shown in Figure 2.6. This has the advantage that the weight of the soil itself stabilizes against its own thrusts. A canal with soil at the back of the training wall demands that it must be calculated alternatively when it is full of water and empty. In any event, the final manufactured volume per lineal metre of training wall is around 0.75 of the square of its total height. It is also same for the trapezoidal cross section.

Sometimes the resistant walls are of rectangular cross sections, with a constant width throughout their height (as with the Júcar Canal). This solution is usually used in excavated sections or placing a support embankment on the outside of the canal.

If the wall is made of reinforced concrete (reinforced with iron bars), Figure 2.7 (*Canal del Piedras*, Huelva), the design techniques for this material must be followed. In general, it is calculated as a single piece formed by the training wall and the bottom. As it is expensive to embed the bottom of the resistant wall in the ground, the training wall is considered to be embedded in the bottom, which transmits the bending moments produced by the soil thrust (empty canal) or the water thrust. If there is soil at the back (the canal is not raised), the water thrust may be calculated by subtracting the active soil thrust from the total thrust. If the land is sufficiently sound rock, it may be considered that it resists the water thrust on its own.

Reinforced concrete retaining walls are much thinner than those made of bulk concrete, but even so they are much thicker than the linings placed over the excavation. The thickness of the reinforced concrete resistant walls is determined to have sufficient width to be able to resist the bending moments well and by the condition of housing the reinforcing rods well. The thickness of the lining alone is determined by the building methods to be used and by the size of the aggregates available or that are economically advisable.

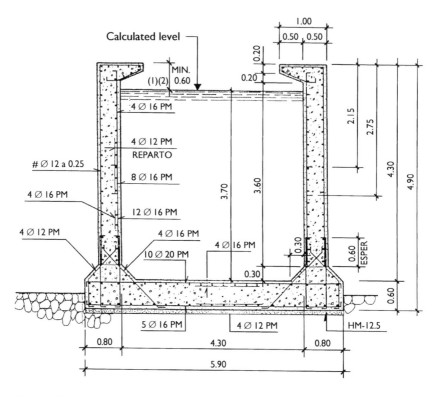

Figure 2.7 Reinforced concrete retaining walls.

As linings are not usually reinforced, the placing of the reinforcing rods does not determine the thickness.

2.7 Possibility of obtaining the desired waterproofness

Under Section 2.2 we fixed 25–50 l/m² in 24 hours as the maximum amount acceptable for the leakage in a canal. The important question that must be answered is this: Is this a value that is easily reached or not, both for unlined canals and for lined canals? It should be considered that an unlined canal and a lined one must be equally waterproof; the difference is that in the first case, the soil itself gives sufficient waterproofness and in the second case we have to resort to a lining.

We should try to find out the capacity that some soils or others have, on their own, to limit the losses to this value and to the contrary, what possibility is there of achieving it using linings.

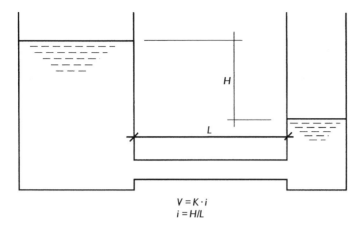

$$V = K \cdot i$$
$$i = H/L$$

Figure 2.8 Darcy's experiment.

We must remember that Darcy's experiment measured the water velocity passing through a tube full of a soil material that joined two communicating vessels with different levels (Figure 2.8). The formula deduced by Darcy is:

$$V = K \cdot i$$

where

    V is the water velocity
    i is the hydraulic gradient (offset between the communicating vessels divided by the length of the tube)
    K is a coefficient, called Darcy permeability.

We must remember that as "i" has no dimension units, the "K" ratio has the dimensions of a velocity. If in the previous formula V is measured in m/s, the Darcy coefficient must come in m/s, although it is more usually measured in cm/s.

If we want to know the maximum permeability that a soil might have in order for a canal built on it to have leaks equivalent to 25 l/m² in 24 hours, even without lining, we will suppose the most unfavourable cross section for the canal. This is obviously a stretch on an embankment, as is shown in Figure 2.9.

In this example a maximum height of the bottom on the land has been supposed to be around 1.5 m, since, in practice, this value rarely is exceeded. The slopes are the ones normally used of the 3/2 type and an upper bank of around 3 m wide has been supposed, which can be used by inspection vehicles or for canals repairs. The maximum distance between the centre of the canal and the outer foot of the slope is 12.25 m.

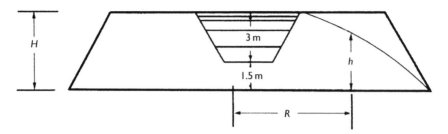

Figure 2.9 Most unfavourable cross section for leakages in a canal.

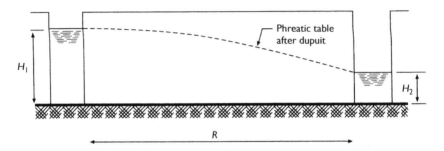

Figure 2.10 Application of Darcy's formula for open drains.

The calculation of the flow leaked by this canal may be made approximately using the formulas deduced from the Darcy formula for studying open drains (Figure 2.10). The formula, as may be seen in many books on hydraulics (36) is:

$$Q = K \cdot L \left(\frac{H^2 - h^2}{2R} \right)$$

in which

    L is the length measured perpendicularly to the paper
    Q is the leaked flow in m³/s
    H, h, R are the measurements indicated in Figure 2.9
    K is the Darcy permeability measured in m/s.

In accordance with the data shown when the canal was similar to the drainage canal, $H = 4\,\text{m}$, $h = 0$, $R = 17.2\,\text{m}$ and $L = 1\,\text{m}$.

In Figure 2.9, the standard canal considered has a maximum wet perimeter of $8.7\,\text{m}^2$ per lineal metre. The total flow for leaks, per lineal metre of canal, will be: $8.7\,\text{m}^2 \times 25\,\text{l/m}^2 = 217.5\,\text{l}$ per 24 hours. The evacuation will be made half on each side of the canal, i.e. $108\,\text{l}$ per 24 hours, or $0.108\,\text{m}^3$ per 24 hours.

Introducing this value as flow into the drainage ditch formula (put in m^3/s), as well as those for H, h, L and R abovementioned, the value for K may be deduced, resulting around $2.5 \times 10^{-6}\,\mathrm{m/s} = 2.5 \times 10^{-4}\,\mathrm{cm/s}$.

The permeability levels of the different soils in accordance with Darcy's formula may be seen in many books (36).

Generally, the following permeability values are accepted:

Gravels and thick sand: from 10^{-3} to $10\,\mathrm{cm/s}$
Fine sands: from 10^{-3} to $10^{-4}\,\mathrm{cm/s}$
Silt: from 10^{-4} to $10^{-6}\,\mathrm{cm/s}$
Clays: from 10^{-6} to $10^{-9}\,\mathrm{cm/s}$
Highly plastic clays: from 10^{-9} to $10^{-10}\,\mathrm{cm/s}$.

The value we have obtained using $2.5 \times 10^{-4}\,\mathrm{cm/s}$ as a soil permeability in which a canal may be built without lining, with losses limited to $25\,\mathrm{l/m^2}$ in 24 hours, is the one corresponding to silts with permeability levels close to the very fine sands.

We may summarize by approximately saying that the silt (or mixtures of soils with equivalent permeability) is the material that separates (regarding to admissible leaks) the soils that allow the construction of unlined canals of those that require a waterproof lining.

If we want to analyse the possibility that a lining on its own manages to obtain leaks lower than 25–$50\,\mathrm{l/m^2}$ in 24 hours, we must consider a canal with a water depth of around $3\,\mathrm{m}$ (in order not to increase the losses due to leakage, irrigation canals very infrequently have depths greater than $3.5\,\mathrm{m}$, however big they are). We can ask what the necessary thickness 'e' necessary would be of a normal clay lining layer that should be placed on top of the excavation of a highly permeable piece of land, in order that a canal built in it would have at most the losses mentioned above.

According to Darcy's Law, the velocity V of water passing through the lining would be given by the expression $V = K \cdot i$, where K is the permeability of the material (in this case the clay) and i is the hydraulic gradient.

For the abovementioned hypothesis we may suppose therefore:

$$K = 10^{-6}\,\mathrm{cm/s}$$

The hydraulic gradient of the leakage will be given by the difference of the inner pressure within the canal (let us say $3\,\mathrm{m}$ high of water) and the pressure outside the canal (null if the land is very permeable and the water table is low, as is normal in most cases), divided by the thickness of the lining, i.e. $300/e$.

If we adopt the maximum admitted leakages in a canal as $25\,\mathrm{l/m^2}$ ($2.5\,\mathrm{cm}$) in 24 hours, the speed with which the water goes through the lining will be:

$$2.5\,\mathrm{cm}/86\,400\,\mathrm{s} = 3 \times 10^{-5}\,\mathrm{cm/s}$$

Applying Darcy's formula, it will be: $3 \times 10^{-5} = 10^{-6} \times 300/e$, from where it may be deduced that this $e = 10\,cm$. In other words, it may be deduced that a 10 cm thick layer of clay acts as a waterproof in the canal in such a way that the maximum leaks will be those desired for 25 l/m^2 in 24 hours.

It has been proved therefore that it is possible to achieve the maximum leakage values that we have proposed.

Under no circumstances may it be deduced that we recommend such a lining; we mention it only in order to fix the magnitude orders for leaks. As may be deduced from the considerations that we will make when dealing with unlined canals (Chapter 11), this covering of only 10 cm thick clay would have a very short life, due to erosion by water velocity.

It is interesting to compare these values with those that certain canals in the world really hold. It is not easy to measure the real losses in existing canals and it is difficult to find references that are worth mentioning. Amongst those that we dare to quote is the following list of canals in the USA (37). On this list in the first column is the name of the canal and the progressive to which the leaks refer. In the second column is the type of lining and in the third, the leaks measured in l/m^2 in 24 hours.

This list should not be used as a basis for statistical studies; rather it shows that there are canals with high leakages and others with very low

Canal	Lining	Losses (l/m^2)
Boise Project, Idaho		
Lateral 0.1–1.0	Asphaltic mix	12
Central Valley, California		
Contra Costa Canal		
1805 + 56 to 1857 + 55	Reinforced concrete	2.6
1857 + 67 to 1873 + 85	Reinforced concrete	27.6
Delta-Mendota Canal		
4355 to 5485	Improved soil	2.6
Courtland Canal		
810 to 820	Soil	45.8
832 to 845	Soil	30.24
Kans Canal, Nebraska		
Helena Valley Canal		
1174 to 1213	Improved soil	242
Meeker-Driftwood Canal		
723 to 738	Soil	15.5
Central Valley		
Friant-Kern Canal	Bulk concrete	21.6
Rio Grande, New Mexico		
Texas West Canal		
51 to 273	Bulk concrete	250
273 to 315	Bulk concrete	153
315 to 343	Bulk concrete	79.5

leakages. But in any event, it is seen that the value that we propose of $25 \, l/m^2$ in 24 hours is attainable.

There is other highly significant data. For example, in the Canal of Las Dehesas (Extremadura, Spain) (87) at the start of the operation of the canal, some significant losses are observed, which are exteriorized at the foot of the embankments.

It was considered that the losses were excessive and a gauging campaign with current meters was carried out all along the stretches of the canal, with the conclusion that the losses were of $59 \, l/m^2$ in 24 hours.

A total revision was carried out of the joints with Thiokol polysulphurate, repeating it after the gauging campaign. The losses had been reduced to $6.8 \, l/m^2$ in 24 hours, which was within the values we are defending and highly acceptable.

Chapter 3

Study and definition of the canal layout and earth movement works to be carried out

3.1 Introduction

Canals have two important characteristics which, although also common to other civil works, are fundamental.

One of them is that they are linear works, of great length, and so they cross soils of diverse categories, sometimes with very different problems that require different local solutions, even having to use cross sections of different shape or dimensions for the canal, adjusting them to the circumstances of each case.

The other fundamental characteristic is that the cost of the work depends in great measure on the adjustment of the work to the field (topography, geotechnics, appropriate selection of machinery, etc.), and so the previous studies, final design and construction should be intimately related.

It is evident that for the study of the layout of the canal the readiness of appropriate topographical plans, supplemented by frequent detailed visits to the site by technical personnel, is essential.

3.2 Basic studies

The scale of the plans to be used depends logically on how advanced the study is and on the difficulties of the area. For definitive project plans it is usually considered necessary to have topographic plans on a scale of 1/2000, with contour lines separated by 0.50 m, or 1 m of separation if the canal is very large. The strip to be surveyed should be selected as a result of rough calculations made upon plans of scale 1/10 000, with contour lines 5 m apart, although it is sometimes necessary to work with the available plans. It is always necessary to see whether there are air stereoscopic photographs, made previously, that allow as detailed a plan as possible to be prepared, although perhaps without field survey.

It is a good recommendation to prepare plans that not only refer strictly to the strip felt to be necessary. Experience demonstrates that it is frequently necessary to make rough calculations of works variations that require larger coverage of topographic plans. If they do not cover the whole area, this may represent a loss of time and supplementary cost. It is especially necessary to foresee the variations that may arise as a consequence of having to include tunnels, flumes, etc.

A general geological study is always very useful, because it overviews the area and points out those areas where a detailed geotechnical study may be necessary.

Also, the geological study may be indispensable for an initial analysis of the tunnels. The study of the flows to be carried out on a canal is usually a result of other different studies, i.e. agronomic for irrigation, energetic for the canals supplying hydroelectric plants, population consumption for water supply. We assume that in the present phase of study of canals the flows are sufficiently defined. The water supply requirements for specific areas may also be a basic project data determining the solution to be chosen.

3.3   First solution fitting-in

With all this and with the hydraulic formulas of a uniform flow, mentioned in the Chapter 1, a first rough calculation can be made and the selection of the cross sections and gradients to be adopted. An initial layout adjusted to the basic conditions demanded to the canal may be chosen taking into account the mild gradients it will have (in general of 10–100 centimetres by Km). Very often the canal seems to be located upon a contour line, separating very little of it, to get the slope selected.

It is the moment to locate the canal approximately on the field and to analyse its characteristics. The classification of soil made by an expert technician, is indispensable as minimum for the method of Casagrande. It is desirable however that these studies come supplemented with granulometric analysis and with the limits of Atterberg. In the opportune moment it will be necessary to get the maximum Proctor densities in the laboratory.

It is necessary to make systematic drills along the canal plan, classifying and also analysing the soils in depth.

The drills will not be generally very deep (except for the study of the flumes foundations or big ancillary works or tunnels), since the excavation depth for the canal will be in general modest, because the canal water depths do not usually exceed 3 or 4 m, however big the canal is. These drills along the channel can be made with light drilling machines, mounted on work vehicles and managed by personnel that do not require a great specialization.

The purpose of these drills is to know the soils that we will find when digging (mainly if there is a possibility that there is rock), being able to foresee the way to use the machinery and fundamentally its production costs. The knowledge of the class of soils and their characteristics will allow to make the necessary studies of stability of banks, drainage, etc. Usually obtaining unaltered soil samples are not necessary for drills along the canal.

The frequency of the drills and the subsequent classification of soils depend mainly on the variability of the soils, which can be appreciated easily in a first approach. In any case, for the definitive project, the drills should not be separated by more than 100 m.

In view of all this and keeping in mind the discussions of previous chapter, we shall be able to decide whether a lining is needed or not, and even its type when required.

In this first adjustment of the canal plan it is also important to detect the affected services which, because its interference with the canal, will influence us in the definitive design. Not only can the highways and railroads demand very important works, but also the electric lines, pipes for water supply, isolated houses or other constructions which can suggest slight plan variations or to require the construction of works of restitution of servitudes.

When, in accordance with the following chapters of this book, the lining to choose, the ancillary works to be inserted in the canal, etc. are defined in detail (or simultaneously to these studies) we can proceed to fix the canal on the land.

3.4 Construction project

Given the mild gradient that in general the canals have, a precise topographic levelling to connect with the national topographic net is needed. It must be extended along the canal. It is necessary to leave level references in landmarks of concrete of enough volume so as to assure their permanency and/or signs in firm rocks, approximately at each kilometre. The landmarks and signs must be located outside the canal strip, so that they are not affected by the future excavation.

The canal axis is materialized on the land by means of stakes placed in all the high or low points and other singular points, at an interval of minimum 50 m. In each one of these points a transverse profile is marked taking out topographical data of its distances and bench marks, taking care that the extreme stakes are also outside the area affected by the works, in order to be able to know in each moment how the primitive land was.

The vertexes of the canal are fundamental points that will be marked in the field by means of references in far away places, so that they are not destroyed during the execution of the excavations. Starting from which you should be able to reconstruct in all moments the situation of the vertex.

The described system is the traditional one that is still the suitable one in many areas where there is no other procedure or when it is more economic, for example in order not to have digitized topographical planes and to be a short canal, or to be coupled to the readiness of appropriate technicians, etc.

If on the contrary there is an aerial survey, with field support to fixed references, it is necessary to reduce field work and increase cabinet work, with the help of computers.

Detailed attention must be given to canal bends which, in order to produce small head losses, should have a minimum radius equal to five times the maximum canal width and in large canals with concrete linings, made with sliding form machines, which we will study in Chapter 5, the minimum radius to be adopted is 20 m. Many times a service road is built on the lateral bench of the canal. Vehicle traffic conditions to be adopted in such case may condition the radius of the canal curves which should be the same as (or very similar to) that of the road.

3.5   Planning of the construction process

We have pointed out the important relationship between the canal design and its execution. Only by foreseeing all phases of the construction process carefully can one make an adequate design and have a sufficiently approximate idea at its final cost.

In several chapters of this book we will study various problems concerning the construction of diverse linings, but there are a series of almost common operations for all canals which we will now analyse. Among them there is a very important group formed by earth works and the construction of access to the different points of the work, to which we will devote special attention now.

The great length of the canals requires that access ways be foreseen for the construction of the different stretches, without neglecting their connection with the large preparatory facilities for materials for the linings, like the batching plants for concrete, production of asphaltic agglomerate, etc. Of course when storing materials along the canal we must take much care not to interfere with future excavation or future works. This means forecasting enough terrain, thinking not only of the final work but also of intermediate situations of construction. The previous expropriation or purchase of the necessary lands is then indispensable.

The location of the concreting facilities, asphaltic mixtures, etc. should be made with a view to the aggregate available and of its location. This topic should be studied attentively, since it can have a great influence on the total cost, the cost of the lining being an imporant chapter in the total budget. In an interesting study conducted on 33 spanish canals (48), all with concrete lining modelled in situ, detailed values are given on different cost aspects. From them we have deduced that approximately, having discounted the budgets for tunnels, aqueducts, flumes, canal-bridges, syphons and other specific works that could mask the values, the final cost for earth movements and concrete could be:

Levelling	31.0%
Canal excavation	10.5%
Excavation trimming	5.3%
Lining	53.0%

The canal excavation refers to the rough excavation of the canal interior that afterwards is trimmed to get the exact section desired, the cost of which per cubic metre is doubtless higher.

The high lining cost, although it may be modified by special circumstances for each canal, doubtless demands a wise selection of its type and construction details.

Sometimes there will be roads or lanes that can be conditioned and other times we shall even have to foresee the construction of new work roads, without ever forgetting the pool problems of lack of drainage that can interrupt the works during the rainy seasons.

Large canals are usually designed with an annexed bank or road running parallel to them. This can be of great utility, but for the first phase of the works it is useless. More effective is the construction, near or alongside the channel, of several provisional lanes that make it possible to connect diverse sites.

Formerly, in concrete lined canals, the first part constructed was the bottom or invert of the cross section, which was used as a track for lorries and works machinery, and also afterwards, with the canal in service, for cleaning operations when the canal was emptied. This may be a comfortable solution if the bottom concrete and the ground that acts as a foundation are able to support the traffic loads, which does not generally happen. This is a decision that should not be taken without a detailed prior analysis. If such a solution is adopted, it is necessary to foresee the construction of access ramps so that the vehicles can enter and leave the channel, as you can see in Figure 3.1 in the A stretch of the Orellana Canal (Spain). These ramps are partly submerged with normal water depths of the canal, but head losses are very small.

It is necessary to foresee (and to require it in the Sheet of Specifications) that before the excavation the contractor must proceed with removal of vegetation and land cleaning. The vegetable matter will be placed in predetermined spots, for use as appropriate.

The surface soils in the area affected by the canal are not fit for use in embankments, because they have seeds and roots of the natural vegetation, therefore the contractor cannot place them underneath any work or the lining. For that it is necessary to make an initial excavation for cleaning and transport the spoil to provisional storing points. It may be a good solution

Figure 3.1 Access ramp in the Orellana Canal.

to level them in areas where ornamental plantations are made, as that may solve ecological problems, improving landscapes, etc.

The cross section of the canal may have different characteristics according to whether it is built in excavation or in embankment.

Some sections are completely dug into the ground. Others are completely built in embankment, but the most normal thing in canals of a medium or large size is that they have a cross section that is partly in excavation and partly in embankment (Figure 3.2).

If the quality of the dug soils so permits it, they should be used in the embankments, but in general their volumes are not compensated, there being an excess of spoils. The habit that has reigned for a long time has been to leave on both sides of the canal formless heaps of earth surpluses that in the classic language were called "spoils".

This habit is not acceptable, for reasons of environmental aesthetics, being advisable to extend the surplus of the spare earth (without necessity of compacting) in a contiguous fringe parallel to the service road of the canal. If for any reason the road or service way is not built, the fringe of extended earth serves at least as a space for walking.

The fringe of extended spare earths should be placed on the lowest part of the cross section and never on the highest part in the hillside, so as not to impede rain water that slips downwards to the canal area. This would be dangerous, because some small pools would form and could produce piping towards the canal excavation. The canal is in fact, jointly with the

Excavated cross section

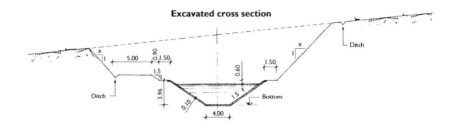

Excavated/Backfilled cross section

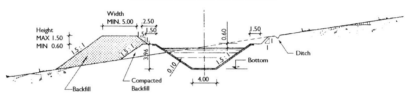

Backfilled cross section

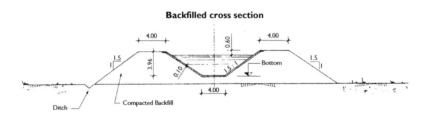

Figure 3.2 Cross sections in excavation and in backfill.

drainage which we will study in Chapter 9, the one responsible for picking up and evacuating the spare waters, before they produce harm, or at least for diminishing their quantity. In Figure 3.3 may be seen a range of spoils that are completely prohibitive.

If the excavations are very important elements in the work of a canal, the embankments are even more important; their possible sliding towards the water can have serious consequences for the canal operation and this could cause damage in the irrigated area. It is an universally accepted norm that their compaction reaches a minimum 95% of density of the Normal Proctor obtained in laboratory. We believe that in the clayey soils (classification A-5, A-6 and A-7 from the A.A.S.H.O.), 100% compaction must be obtained. This requires that the compaction be made by layers of 25 cm maximum

Figure 3.3 Prohibitive range of spoils.

of compacted thickness, with the needed wetting to obtain the humidity for the maximum Proctor density secured in laboratory with the adequate machinery and the number of passings needed. The compaction control of all the layers is essential, either by the traditional method of taking of samples and obtention of volume and weight of the dry sample, or by the use of radioactive isotopes.

There are two systems of constructing the mixed cross sections, half on excavation and half on embankment. One consists in obtaining a complete levelling of the combined canal/road first, with the necessary compaction in the embankment part, and then digging the canal section. The other one consists in trying to avoid the dual operation that sometimes occurs, of making an embankment first in an area where the canal is dug later. For this the necessary areas are excavated and at the same time, with the earth thus obtained, another area is earth-filled and compacted. This system can only be used with canals larger than a certain size.

It is very important to analyse the climatology of the area previously, in order to foresee the times of possible work, in which the rain or other meteorological phenomena do not impede the works (for example because the compaction of embankments it is impossible when there is excessive humidity in the land). Only this way will it be possible to study a plan of work adjusted to reality and to come up with an appropriate design for the canal.

If the canal has sufficiently large dimensions, the excavation and compaction of the lateral embankment should be done with some scrapers or with crawler-type scrapers that work in closed circuits, digging inside the canal and transporting and distributing the earth in thin layers to the embankment, at the same time that they compact it, taking advantage of their own weight. This compaction may or may not be supplemented by the use of some compactor rollers which should be of sheep-foot roller type in clayey soils and tire type in other soils.

For spreading and smoothing the extended earth, moto graders are sometimes very useful.

If the dimensions of the canal do not allow the use of scrapers, it is necessary to foresee two types of machines: a back hoe (under wet conditions sometimes a dragline) to make the excavation from one or both banks (depending on the canal width) and bull dozers to extend and compact the embankment by layers. In such cases. the excavation becomes independent to some extent of the compaction of the embankment. When compacting the embankment, the slopes must be made with an increase of about 60 cm, as the lateral surface is always poorly compacted, to drag and trim the excess later.

If the ground is dump, the excavation may require a clamshell or perhaps a drag line.

In Figure 3.4 an outline of the machinery more commonly employed in earth movement works for canal construction may be seen.

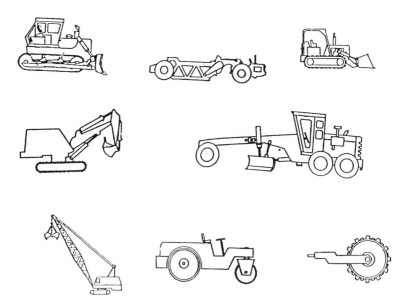

Figure 3.4 Outline of machinery employed for canal construction.

It should not be forgotten that a canal of some medium or large dimensions, once built, is an almost impassable barrier for vehicles and pedestrian. From the beginning of execution of the works, concern for the construction of crossings on roads and existing sidewalks should be borne in mind, before they are cut by the canal. This concern should also be had during the planning and design of the works, because surely vehicles traffic will increase once the canal is in service, and an increase in the number of crossings above the canal will be needed.

No matter how carefully the excavation is made, there will always be errors that are admissible within certain limits that are called "tolerances". Once the trimming of the excavation is carried out and before placing the lining if any, the maximum acceptable errors horizontally, with reference to the theoretical line, are 2 cm in straight tracts, 5 cm in tangents to the curves and 10 cm in full curve.

This requires that the ground be easily dug because if it is rock or an intermediate material between earth and rock, this obviously cannot be achieved. In this case the designer must foresee stuffing the holes with poor concrete (of low cement dosage), well compacted, since filling them with compacted soil will not succeed.

In the canal construction some water must be added to soils when needed in order to get the maximum Proctor density obtained in laboratory and for concrete production. The engineer must have concern over how he or she will get the water supply.

At this time we recall what was said in epigraph 2.2, where we said that the canal should have the smallest number possible of swelling/drying cycles. If the canal is constructed from its main intake in a downstream direction, both problems are solved. Nevertheless, due to the short time available for construction, the works must begin simultaneously in various stretches.

If problems of dampness and high water tables exist, the first operation to carry out is to drain the water and to get the lands dry, since the compaction of the soil, in order to get high densities that are representative of high resistance, should always be done with optimum Proctor humidity.

Drainage channels, given their very purpose, are usually made in low areas, where normally there is dampness and even superficial currents.

The elimination or deviation of the superficial and phreatic waters in a drainage channel is a basic problem that must be solved. Frequently the excavation must begin from the outlet in upstream direction, sending water downstream, taking advantage of the constructed stretches.

Many a time the drainage canals must have great depth, due to the topographic conditions, which causes their cross section to be bigger than that demanded by the flow to be carried. If the canal is made with a trapezoidal cross section, a narrow base (for example 1 m wide) is often selected, and the slopes are chosen according to the ground stability.

Chapter 4

Mass concrete lining

4.1 The suitability of concrete as a canal lining

Concrete is a product obtained through the intimate mixture of gravel, fine gravel and sand (in suitable proportions), together with cement and water. Either after or simultaneously with the pouring of the paste into the mould or formwork that provides the shape of the part to be manufactured, it is essential that the mixture is correctly compacted in order to reduce as much as possible any interior cavities. Following this, there is a process called "setting", during which the mixture hardens, heat is given off and the manufactured part undergoes a reduction in dimensions known as "shrinkage".

Cement concrete is very frequently employed as a canal lining material. However, the conditions to which it is exposed are extremely difficult and, to a certain extent, are not suited to its characteristics.

It must be remembered that the Bureau of Reclamation, in its book on concrete, considered that concrete employed in canal linings is the one that has to withstand the most arduous of conditions (9). For this reason, it recommends the use of air entraining agents, a study of the water–cement ratio, the use of perfect grain size, very good compaction and that close attention be paid during the concrete pouring operation, etc.

Cement concrete is extremely impermeable and only small amounts of filtration would occur if it were not for the very great danger of cracking involved. This is because, although it is a material with excellent compressive strength, it has, on the other hand, poor tensile strength. It is a highly rigid material that adapts very poorly to ground strain, no matter how small they are because the material is quite unable to withstand them. Significant stress is also produced due to shrinkage and temperature drops, which enormously increase the danger of cracking appearing, which then becomes the cause of serious filtration.

Water loss in mass concrete-lined canals does not occur in a uniform fashion over the entire lining, but instead is localized to the cracks or poorly impermeabilized joints.

For this reason then, it is essential that adequate, correctly imperme-abilized joints are employed in canals that are lined with this material so that any lining movement takes place precisely via these joints, preventing cracking and hence there will be no filtration.

4.2 Study of joints in concrete-lined canals

The joints may be either transversal or longitudinal to the canal. Figure 4.1 shows the Provence Canal in France, in which the transversal and longitu-dinal joints can be seen. There are also construction joints.

The transversal joints may be either contraction or expansion joints. The purpose of the former is to prevent cracking caused by concrete shrinkage or contraction due to reductions in temperature.

Transversal expansion joints are employed to prevent the problems caused by the heat expansion in the lining.

The longitudinal joints are intended to provide the canal with a type of articulation which, to a certain extent, will allow it to follow strains in the ground caused by changes in humidity levels or differences in settling in nearby areas of different compaction.

The main purpose of the construction joint is to provide adequate ter-mination of the lining completed during one day's work so that it can be correctly joined to the next day's work and so on, and also to join new concrete sections to older ones.

Each of the different types of joints will be examined below.

Figure 4.1 Joints in the Provence Canal (France).

4.2.1 Construction joints

A characteristic of concrete is that when recently made, it does not join to the previous, already set concrete. It results in a highly unfavourable situation where this material is being employed as a lining. This problem leads to the requirements for construction joints. These are typically used to join one day's construction work with that of the previous day.

There is very little we can say at this time about the separation of construction joints. They are conditioned by project specifications, construction methods and execution speed. It can only be pointed out that for construction purposes they should be the same, or at least similar to the other described joints.

Construction joints are also sometimes needed in the longitudinal direction, in this case, not to join one day's work with that of the next, but to tie together the work carried out in various stages. One very good example of this is in the construction of a canal where the floor is executed first, followed by the slopes. (In this book the term "floor" is used to indicate the concrete base or bottom of the canal and the "slopes" are the concrete sides.)

A construction joint, which, by the way, is extremely dangerous, must be made between these two elements. The fact is this joint is not impermeabilized in many cases and since the most recent concrete does not unite with the older sections, it becomes the source of significant leaks because the pressure at this point corresponds to the maximum canal depth. Moreover, the joint between the floor and walls is usually defective due to the included angle formed by these elements.

In addition to this, when the canal walls are concreted with the old method of manufacturing the concrete at the highest part of the canal and then pouring it down the slope, the largest size gravel separates from the rest and falls first so that a type of concrete without fines, which is of extremely poor quality, with little grout, forms at the bottom of the slope. This joint must be treated and studied with the utmost care (Figure 4.2).

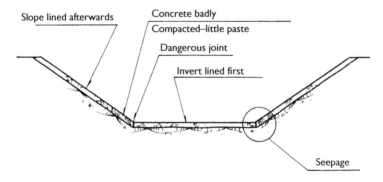

Figure 4.2 Joint between floor and slopes.

Among the measures that may be adopted, we will first describe that of concreting a complete canal section in one operation so that there is no longer any requirement for this type of joint. If this is not possible, then it is necessary that the concreting of the lining that is to be joined a previously concreted section must commence with a very careful cleaning of the old concrete, together with the preparation of an adequately impermeabilized longitudinal joint. In addition to this, it is recommended that if the concreting of the floor is carried out before that of the walls, it should also include the lower sections of the walls so that the construction joint is not located at the floor-wall angle, but instead on a flat slope section. This will avoid the always-dangerous floor-wall angle (Figure 4.3).

4.2.2 Longitudinal joints

On many occasions, longitudinal joints are included in canal linings to prevent fracturing caused by transverse section slope movement. It is quite obvious that what is usually referred to as the "canal lining", which is only a few centimetres thick, cannot possibly withstand ground pressure. This requires a much greater thickness of concrete in what are called "lateral walls".

If ground movement occurs, then the canal lining will most certainly crack. In order to prevent this, longitudinal joints are sometimes employed as a preventive measure, which act as articulated joints that allow localized lining movement, producing what could be described as an articulated lining.

Ground movement is occasionally produced by clay swelling, which, in certain situations, can lead to significant pressure resulting from changes in moisture levels. However, the most frequent case is produced by differential settling of the various soil layers, most often in the situation of a poorly

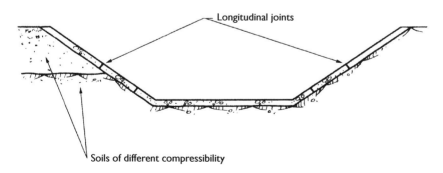

Figure 4.3 Sketch of longitudinal joints.

compacted backfill located on top of natural ground. In all these cases, where it is possible to predict different behaviour of the various layers, it is recommended to employ longitudinal joints in suitable locations. This is especially applicable to the contact zone between the natural ground, which has become compacted with the passage of the centuries, and the backfill constructed during the works, unless, of course, very strict compaction operations were carried out during its execution.

Since floor settling is a mainly vertical movement and wall settling typically involves turning of the lining, it is recommended that one of the longitudinal joints is located low down in the walls, close to the floor.

The importance of correct compaction for backfilling and the densities that are to be obtained have already been pointed out in Chapter 3. The amount of settlement is reduced with good levels of compaction and hence, the danger of lining cracking.

Soil swelling, which is invariably caused by changes in moisture level, requires special attention. The severest conditions occur when the canal is first filled, unless special precautions are taken to prevent this, such as filling the canal with water as construction advances. Then the soil changes from maximum dryness to maximum wetness and usually produces the first and most serious cracking.

It may be recommendable that the soil compaction be performed, not with the optimum level of moisture, but with one that is slightly higher. Since the obtained soil moisture/dry density curve is at a maximum with a horizontal tangent, a small percentage moisture variation will hardly reduce the dry density. The canal should be filled with water just as soon as possible, before it is put into service, in order to maintain as constant a moisture level as possible. This is obtained almost automatically in many cases during normal operation.

However, most cases of horizontal cracking found in the concrete linings of canals are due to the fact that the slopes are too steep to be stable. Although at first sight, it would appear that certain slopes are stable, their behaviour is, in fact, deficient because of the extreme wet/dry conditions to which the canal lining extrados is subjected. For this reason, it is not recommended that the canal slopes be made at angles that are greater than those corresponding to an inclination of 3 in base and 2 in height, except where, for other reasons, together with providing sustaining strength, the walls are constructed at an angle steeper than 45°. It is not recommended to construct walls close to an inclination of 1:1, since in these situations, the soil is not usually stable, and the concreting operation is neither easy nor economic since formwork has to be employed because the soft concrete will have a tendency to flow downwards. This point, with reference to lining manufacture, will be covered in more detail in Chapter 5.

The movement produced in canal slope linings due to slope instability normally leads to more or less horizontal cracking at between one-third and one-half of the height from the floor. Under circumstances where cracking has occurred because the precaution of constructing suitable slope angles was not adopted, then the easiest repair method consists of enlarging the cracks and employ a correctly impermeabilized joint.

Figure 4.4 shows the damage to an unfinished canal that was caused by heavy rains just after constructional operations that were carried out during very dry weather, which caused ground swelling, rotation of the lined slopes, which separated from the soil and consequently producing of longitudinal cracks.

It is quite evident that this canal's slopes were too steep (at 45°) and also suffered from drainage problems (see Chapter 9), because the excavated soil was piled up without any thought given to the evacuation of the water running down the side, which collected behind leading to ground filtration and its immediate swelling.

However, in spite of the fact that stable slopes are constructed and attention is paid to drainage problems, it is still quite normal for longitudinal cracks to appear at a height between one-third and one-half of the canal depth as shown in Figure 4.5 which shows the Calagua Canal in Bella-Union (Uruguay). This clearly shows the need for longitudinal joints.

Figure 4.4 Ground swelling – Orellana secondary canal.

Figure 4.5 Calagua Canal in Bella Union (Uruguay).

4.2.3 Acceptable movement between transversal contraction joints: Predicted movements

As has already been explained, the purpose of transverse contraction joints is to prevent the consequences of concrete contraction cause by drops in temperature and setting shrinkage.

The distance between these transverse contraction joints is determined by the condition that the concrete traction forces are less than those it can withstand.

These traction forces are produced by the sliding friction between the concrete lining and the underlying soil. This phenomenon may be better understood by imagining what happens as follows: first, due either to a fall in temperature or setting shrinkage, the concrete contracts. Secondly, the bonding forces between the soil and the lining oppose this strain and stretch it so that it returns to the original situation by subjecting it to traction.

The maximum traction forces, and hence the resulting cracking, appear precisely at a point equidistant between two consecutive joints. The reasons are quite clear. On the one hand, due to symmetry reasons, this must be at the centre. On the other, this is the point where the traction forces are greatest, since the linkage forces between soil and lining increase from the end of the slab towards the centre as demonstrated in Figure 4.6.

If the bonding force between the soil and the lining is F kg/cm^2, which opposes the sliding, at a distance of L cm between two consecutive contraction joints, with e being the lining thickness in cm and t kg/cm^2 is the acceptable traction in the concrete, at the moment of fracture the traction force acting on each side of the central section is given by $(F \times L)/2$.

This force acts in the contact plane between the soil and the lining and produces a bending moment in the concrete of:

$$\frac{F \times L}{2} \times \frac{e}{2} = \frac{F \cdot L \cdot e}{4}$$

The resisting moment of the rectangular section will be $e^2/6$ and the total traction, which is the sum of the pure traction plus that due to the bending moment, is given by:

$$t = \frac{F \cdot L}{2 \cdot e} + \frac{F \cdot L \cdot e}{4} \times \frac{6}{e^2} = \frac{2 \cdot F \cdot L}{e}$$

From this formula, which is the basic one for selecting the transverse contraction joint separation, we can calculate the value of L (joint separation) by means of the following equation:

$$L = \frac{t \cdot e}{2 \cdot F}$$

provided the other values are known. There is no problem in establishing the thickness or the tensile strength t, using the breakage of test pieces or other suitable system. The real difficulty lies in the establishment of F, which is the result of two different addends.

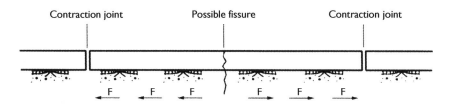

Figure 4.6 Location of contraction joints.

The first of these is the soil's cohesion value and depends on changes in moisture levels, while the second is produced by the angle of friction and is proportional to the normal component that acts by pressing the lining against the soil.

These values can vary quite considerably from one area of soil to another and hence, F will also vary. However, the worst part is that its value depends mainly on the moisture level of the lining extrados and on the degree of termination of the excavation (also known as "trimming"); in other words, the degree of union between the lining and the ground.

Consequently, only on few occasions is it possible to calculate the value of L, as we would have liked. However, there are many observations and much experience available with respect to real canals that recommend contraction joints being placed at intervals of between 3 and 4.5 m, depending on the lining thickness (25).

However, the previously calculated expression can lead to important consequences and teachings. In a first approximation, it may be considered that the required separation between joints is proportional to the lining thickness, which means that the thinner the lining, the closer the joints have to be. The first highly significant conclusion is that canal linings should not be constructed using thin mortar (for example, with gunite). This would require absurdly close joints.

Similarly, from the previous formula, it can be stated that the better the quality of the lining, with its higher value of t, the greater the separation between the joints, L.

Since the greater the value of F, the smaller the L may be, we can state that the more perfect the excavation profile, the greater the contraction joint separation can be than in the case of poor termination.

All this means that in cases such as concrete linings over an excavation of poorly cut rock, employing contraction joints would be quite useless because their separation would be very small. Fortunately, in these situations there are normally no problems associated with filtration through the cracks.

No less important is the conclusion that may be obtained from the formula by which the separation for transverse contraction joints does not depend on the environmental temperature and hence not on the weather conditions for where the canal is located either. On the other hand, weather conditions do have great influence on the amplitude of joint width variation.

The predictable movements in contraction joints are given below. Thermal dilation for concrete is given by

$$(T \times L)/10^{-5}$$

where T is the thermal variation in centigrade and L is the joint separation in millimetres. This means that for a joint separation of 4 m, 4000 mm, together with a temperature variation of 40°, the joint movement would be:

$$\frac{40 \times 4000}{10^5} = 1.6 \, \text{mm}$$

However, it must be taken into account that these values may be increased if the concreting operations take place at maximum temperatures and that the shrinkage is added to these. The Structural Concrete Instruction (77) establishes that, for calculation purposes, setting shrinkage should be taken as being in the order of 1/4 mm per metre length of concrete. This is the equivalent of a 25° drop in temperature, which when added to the shrinkage due to temperature, assuming that the canal is concreted at maximum temperature, would increase strain levels by approximately 50%.

Quite the opposite effect is produced when the concreting operations are carried out during low environmental temperatures. The thermal effect is highly compensated by the setting shrinkage in the case of temperature rises.

The sole purpose of these considerations is to establish an order of magnitude for the maximum joint strain. It is in no way an attempt to defend the concreting of canals at either extremely high or low temperatures, since the Concrete Instruction and good practices must be followed in all cases.

The following may be listed as a summary for the basic specifications for transverse contraction joints employed in canal construction:

1   They should be placed at fixed distances depending on the lining thickness, the concrete quality and degree of perfection employed in excavation termination.
2   In practice, this separation normally lies between 3 and 4 m (occasionally 4.5 m) and does not depend on temperature variations at the canal's location.
3   The joint must be designed so that it can open (which is the same as allowing a shortening of the concrete between two consecutive joints) by a maximum of approximately 1.5 mm. This value does depend on the site's maximum temperature variation.

It would be an interesting exercise to compare these specifications with those that are now going to be calculated for expansion joints.

4.2.4 Transverse expansion joints: Their location and expected movements

It has already been stated that the purpose of transverse expansion joints is to overcome the expansion problems that occur in concrete when the temperature rises.

There are many conflicting opinions with respect to the use of these joints, with those who defend their location at a distance of 20 m and others who propose much increased separation and even their complete elimination.

Figure 4.7 shows a privately owned brick canal on arcades in the region of the River Henares (Madrid), which has ruptured by expansion due to rising temperatures. Not shown in the figure, on the right side, is a syphon. This provides very clear proof that temperature increases can be very damaging.

The expansion forces caused by this phenomenon lead to a situation of general compression in the canal lining. If it is taken into consideration that the concrete lining of a canal is normally very thin, between 8 and 15 cm, then, since its length produces exceptional slenderness, it can be assumed to be subject to buckling. Fortunately, the lining's own weight and its welding to the ground contribute high levels of assistance in combating any buckling.

Experience reveals that very few concrete linings of canals are damaged by compression forces. There are several reasons here that play in our favour.

First of all, the concrete shrinkage phenomenon plays an extremely important and favourable role with respect to expansion forces within the concrete.

Because of the concrete setting shrinkage, which is equivalent to the contraction produced by a 25° fall in temperature, the effect of a 40° temperature rise is greatly reduced to less than half. If it is then taken into account that concreting operations are not normally carried out at low temperatures, the expansion produced up to the maximum expected temperatures is even smaller. Because of this, the compression stress caused

Figure 4.7 Tertiary canal ruptured by expansion.

by the concrete expansion, assuming that this is completely free from either elongation or contraction, will be much less than those initially expected.

Assuming a concrete modulus of elasticity of $200\,000\,\text{kg/cm}^2$, together with a temperature variation that reaches a maximum of $50°$, remembering that the unit thermal strain of concrete is $1/100\,000$ per degree, then the maximum compression that could be produced in the canal's concrete lining, assuming that elongation is completely avoided and by simply applying Hooke's Law, is given by:

$$200\,000\,\text{kg/cm}^2 \times 1/100\,000 \times (50 - 25)° = 50\,\text{kg/cm}^2$$

This value of forces can be quite easily withstood by normal concrete. However, the fact is that these values are not even reached because of the favourable effect of the contraction joints, which have opened due to the fall in temperature and which will close again as temperature rises once more. Moreover, the reality of the situation is that the maximum temperature increase to be considered is not the maximum expected for the area, but instead the difference between the maximum temperature and that existing during the concreting operations. Since the regulations in all countries stipulate that concrete should not be poured under cold conditions close to freezing, the maximum expansion force in the concrete will be less than the value just calculated.

In consequence, in the actual canal itself, the compression forces caused by rising temperatures are withstood by the concrete and, therefore, expansion joints are not required.

However, we have frequently observed that the syphon inlets and outlets, which are inserted into long section of ditches and canals, become cracked halfway up due precisely to lining pressure, which is under compression. This phenomenon is exactly what happened in Figure 4.7. This is evident proof of the fact that although the concrete lining, or one of other material such as brick, is subject to compression forces that it is able to quite easily withstand, it is also subject to compression forces that may be transmitted to other elements inserted in the canal that are unable to resist them.

It may therefore be established that expansion joints are essential at all points where compression forces acting on the lining could produce certain classes of damage to inserted concrete elements. Special emphasis may be placed on connections with syphon inlets and outlets, with flumes and even before tight curves, even more especially if these are located along sections of backfill. In fact, the lining expansion in two straight lines, joined by a tight curve, can produce radial pressure that moves the curve in spite of soil resistance, leading to filtration problems.

In accordance with these ideas, there is no point in employing expansion joints, as has been carried out in the construction of several canals, with separations of 20 m, or any other fixed distance.

This same position is defended by the International Commission for Irrigation and Drainage, which has stated that "expansion joints are not necessary in concrete linings, except when fixed structures exist in the middle of the canal".

It should also be remembered that expansion joints are elements that are both difficult and expensive to construct, especially in view of the modern concreting techniques, currently employed in canal construction, using longitudinal slip formwork, which makes it easy to produce contraction joints, but not expansion joints due their different morphology.

Once the locations for the expansion joints have been established, the movements they will have to withstand must be predicted so that an adequate design can be prepared.

The problem is complex. If the distances between the joints are large, the part of lining furthest away will suffer practically no movement because the soil itself will apply pressure to hold it in place. In addition, the sections situated to the left and the right at any point along the canal will also collaborate in preventing movement. On the other hand, a slipping of the lining over the soil will occur in the areas close to the joints. It is precisely the strain of the slipping part that causes the movement of the joint.

The phenomenon is complicated because it is internally hyperstatic and involves interrelated strain of both the lining and the soil.

The associated mathematics have not been fully resolved yet; however, it is possible to perform approximate calculations and, above all, to make certain important considerations.

Figure 4.8 contains a drawing of how movement is produced at an expansion joint. The general canal section does not suffer any movement because it is anchored not only by the soil, but also by the compression forces from the anterior and the posterior sections, which are equal due to symmetry. However, the area close to the expansion joint moves when the canal expands, and will partially close the joint. There are both friction and cohesion forces between the lining and the soil.

There is a transition zone between the slip area and the anchored area, in which there is no slipping between the soil and the lining, but there is,

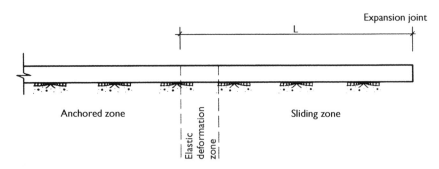

Figure 4.8 Lining movement at the expansion joint.

however, combined elastic strain of the soil and the lining. This area, which is more difficult to study, is much shorter and has much less influence on the phenomenon than the transition zone, and for this reason it will not be taken into account.

Where

L is the length of canal that slips, in cm
P is the perimeter of the canal's transverse section, in cm
e is the lining thickness, in cm
F is the assumed uniform coercive force from the soil that acts on the lining, in kg/cm²
T is the temperature increase, in degrees centigrade
E is the lining material's modulus of elasticity, in kg/cm²
t is the compression produced in the concrete, in kg/cm²
dL is the increase in section length, which is equal to the variation in joint width.

Remembering that the thermal strain coefficient for concrete is 1/100 000, when Hooke's Law is applied to that part of the canal that is described as being anchored because it does not suffer any translation or strain, we will obtain:

$$t(kg/cm^2)/E(kg/cm^2) = T/100\,000$$

from which the compression in the concrete can be calculated as:

$$t = \frac{E \cdot T}{100\,000}$$

The force of the pressure in the anchored area over the slipping area will be:

$(E \times T \times P \times e)/100\,000$ and the force opposing the slipping section will be $L \cdot P \cdot F$.

Both forces must be equal (at the end of the strain), so that:

$$(E \times T \times P \times e)/100\,000 = L \times P \times F$$

The length of the section that slips may be calculated from this equation, which is:

$$L = \frac{E \times T \times e}{100\,000 F}$$

Along the entire length L of the slipping section, the concrete compression reduces linearly, because the bonding force with the soil reduces linearly, from the maximum at the beginning of the section to a value of zero at the end since it is assumed it no longer pushes the joint. The unit strains decrease in a similar fashion and the movement or total shift at the end will be, as in all laws that linearly diminish, half of that if the compression had been constant and equal to the maximum value, as follows:

$$dL = 0.5 \times L \times T/100\,000$$

and substituting the obtained value for L, becomes:

$$dL = 0.5 \times E \times e(T/100\,000)^2/F$$

From these calculations, it can be established that the length of canal near the expansion joint, which undergoes translation, is proportional to the lining's modulus of elasticity, the lining thickness and the square increase in temperature in degrees centigrade, and is inversely proportional to the coercive force of the soil, which means that it is affected by degree of quality in the excavation termination, which is normally called "excavation trim".

The closing of the expansion joint is, in turn, proportional to the modulus of elasticity and the thickness, and inversely proportional to the soil's coercive force. However, it has to be said that the influence of the temperature increase on the closing of the joint is very great, because in the formula, the value of T is raised to the square power and the closing due to temperature is proportional to this value.

It would be an interesting exercise to establish the order of magnitude that these values have in practice. As an example, if we consider $E = 200\,000\,\text{kg/cm}^2$, $T = 40°$, $e = 15\,\text{cm}$ and $F = 0.2\,\text{kg/cm}^2$ (this, of course, may take on quite different values), the result is that the length affected by the translation is approximately 62 m, and the joint closes by about 1.25 cm.

These results largely agree with those obtained in practice, in which it can be seen that the expected expansion joint strain is approximately 1 cm, an amount that on occasions will be larger.

As a basic summing up exercise of an expansion joint's properties, it may be compared with those of a contraction joint in order to make their large differences quite clear, as follows:

1   Expansion joints should only be located before and after specific works inserted in the canal in order to free them from canal pressure.
2   The strain expected for the expansion joint is a minimum of 1 cm, an amount that on occasions will be larger.
3   The lining thickness, its modulus of elasticity and the excavation termination quality, all influence the expected expansion strain. However, it is most affected by temperature rises.

Once armed with this information, we are in a position to design the contraction and expansion joints, which will obviously be different in view of their widely different characteristics.

4.3 Forces that must be supported by the joints: Joint models

The materials employed in the construction of expansion joints are required to work under extreme conditions. Above all, they have to be capable of withstanding strain that, although with apparently small absolute values, is enormously large in value with respect to their thickness. The accepted repeated elongation in the joint material is a fundamental piece of information for its design.

The expected strain in the contraction or expansion joints that we have accepted, together with shape considerations, which will be examined later, mean that the impermeabilization must be able to withstand strain that may be greater than 50% of its original dimension. This condition assumes that the adherence between the joint material and the contiguous piece of concrete is sufficiently large to prevent any loosening and subsequent filtration. Experience has demonstrated that in most cases this is the mechanism that leads to joint failure.

In canal, the transverse joint is subject to loosening of the impermeabilization material due to the effect of high temperatures. This has a highly unfavourable effect and has been the cause of innumerable failures.

The described conditions are even more difficult if it is taken into account that the material has to withstand not only the low temperatures, but also the highest expected ones. The temperature swings that have to be considered are higher for the joint materials than for the actual concrete because the joints, which are often black, will have a greater range of variation. It must often be taken into account that the joints have to withstand temperatures of $+60$ or even $+70\,°C$.

The low temperatures require that the joint material does not become brittle or fragile, that it does not peel away from the concrete and that any cracks that do form are not deep enough to endanger the overall impermeability.

One other effect that the joints have to withstand is ageing, which may be accelerated by the ultra violet radiation in sunlight and the absorption of water.

The design of the various types of joints must pay very careful attention to all the previously described points.

Since the conditioning factors for contraction joints are quite different to those for expansion joints, the design models to be employed in each case are also quite different. We will now take a look at the various possibilities of joint design, first for contraction and then for expansion.

4.3.1   Models for transverse contraction joints

Previously, we saw that contraction joints must have the possibility to open up a maximum of 1.5 mm for temperature drops or concrete setting shrinkage. It does not matter that the increase in temperature causes the two sides of the concrete to come into contact with each other because the expansion joints, if they are required, are employed to combat this effect.

One frequently used type of joint in Spain for canals concreted by traditional methods (Section 5.1) can be seen in Figure 4.9. With this system, concrete sills were constructed in trenches dug in the soil inside the future canal so that the outer wall exactly coincides with the plane of the required profiled excavation. The concrete lining in the form of slabs separated by the joint located on the beam axis is then supported on these beams. In Spain, these slabs are known as "cloths" because they remind one of a piece of fabric supported on the excavation.

Alternate slabs are constructed first (Figure 4.10) and once they have set and undergone most concrete shrinkage, the joints are painted with impermeabilization products and then the intermediate closure slabs are concreted. Use of this system reduces shrinkage effects to almost half, so that the joint apertures are much smaller.

One disadvantage of this system is that although the impermeabilization product adheres perfectly to the oldest slab, this is not the case for the

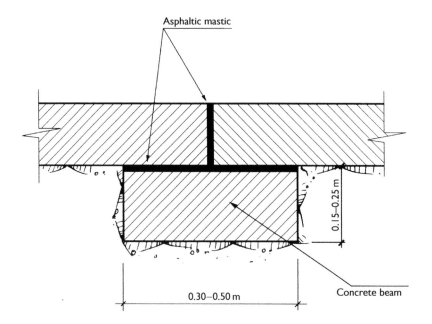

Figure 4.9 Old construction joints.

Figure 4.10 Construction of alternate slabs.

fresh concrete used for the closure slabs. This obviously casts doubt on the quality of the impermeabilization operation, although they are, in general, efficient.

One type of widely employed contraction joint, especially for floors (invert), is shown in Figure 4.11. It does, however, involve a certain danger of breaking because of the thin amount of concrete.

Typical contraction joints consist of a slot or the beginning of an artificially produced crack, which is suitably impermeabilized and continued by means of another crack that fully penetrates the thickness of the concrete, which is produced by the concrete when it contracts, Figure 4.12.

This type of joint is very suitable for canals that are concreted in a continuous manner. This system of execution consists of cutting a slot in either the set or the fresh concrete. The former case requires special disk saws, whereas the latter can be accomplished using suitable blades.

The technical literature usually recommends that these slots are approximately one-third of the concrete thickness.

This is not, however, very easy to accomplish. It must be remembered that concrete includes thick elements, the lining of which is limited to 25% of the piece's thickness by the regulations of most countries (77). Nevertheless, it is normal for small thicknesses, such as in concrete piping or canal lining, for the associated Standards to permit a gravel diameter of up to 33% of the thickness.

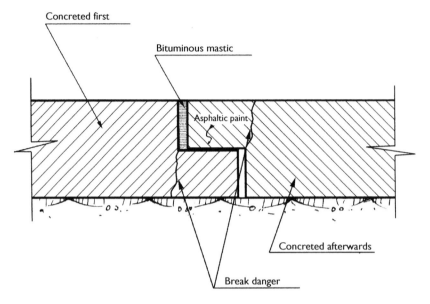

Figure 4.11 Stepped construction joint.

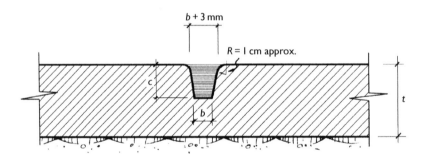

Figure 4.12 Contraction joint.

If the joint is produced employing a blade in the fresh concrete, which leaves a slot of approximately 33% of the total thickness, this could cause displacement of the larger aggregates leading to a breaking up of the actual concrete structure.

If the joint is made in fully set concrete, then the structure will not be disturbed, but it does require cutting of the aggregates and this type of operation involves the use of costly, specialized machinery.

The best solution is to make the joints using fresh concrete followed by re-compaction of the joint sides if the aggregates are moved.

The table, which is now considered a classic, provided by the Bureau of Reclamation (25) and reproduced by the International Commission for Irrigation and Drainage (56), containing the dimensions for the described joint is given below.

t (cm)	b (mm)	c (mm)	Approximate separation between joints
5.08	6.35–9.53	15.88–19.00	3.048
6.35	6.35–9.53	19.90–22.23	3.048
7.62	9.53–12.7	25.4–28.58	3.66–4.57
8.89	9.53–12.7	28.58–31.75	3.66–4.57
10.16	9.53–12.7	31.75–34.93	3.66–4.57

The described joint is very suitable for transverse contraction joints.

Several different types of contraction joints have been used, adapted to the various types of canal. These all follow the same principle in which the joint is able to open a small amount of less than 1.5 mm and where both sides can come into contact with each other if the temperature rises.

Included in these are the joints designed for self-standing side-walls in flumes or lightweight sections that basically consist of a hole that is approximately parallel to the facing and which is filled with clay once the walls have contracted (Figure 4.13). The contact surfaces between two consecutive walls are painted with an impermeabilization product. The main advantage of this type of joint is that they can be repaired with relative ease by replacing and compacting the clay from the upper section in the case of filtration.

Its main defect lies in the difficulty of correctly compacting the lower section of the clay, which means that the diameter must be sufficiently large and great care must be taken not to allow large elements to become trapped inside.

In this case, there is no danger of the impermeabilization materials loosening because of the thinness of the joint. As with the previous case, it cannot be expected that the fresh concrete will adhere perfectly to the old impermeabilization material and it can be even less expected that the thin layer will be capable of withstanding the strain imposed by the slabs. When the walls are of significant size, then this type of joint can be constructed employing a larger section and hence improving its efficiency. An example of this is the joint we employed for the El Burro Flume on the Orellana Canal and which is shown in Figure 4.14. Figure 4.15 shows this flume.

4.3.2 Longitudinal joints

A new type of contraction joint is being employed for longitudinal joints that consists of a strip of prefabricated flexible material, normally PVC, in

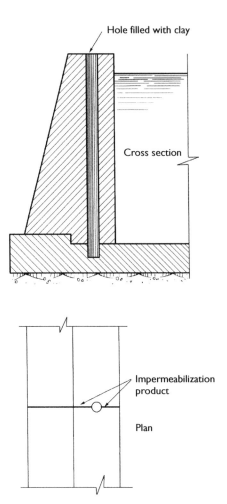

Figure 4.13 Joint for lateral walls of canals.

the shape indicated in Figure 4.16 and which is inserted into the concrete as it is being spread.

The finished joint is shown in Figure 4.17. The philosophy behind this joint is the same as that of the previously described one, in which a strip of plastic produces a localized area of weakening so that any cracking will occur precisely at this point. The shape of the plastic strip forces any possible filtration to take a long path, which means that any losses are small.

The machine that lays the plastic strip also vibrates the concrete, which will ensure a high level of coherence between the two.

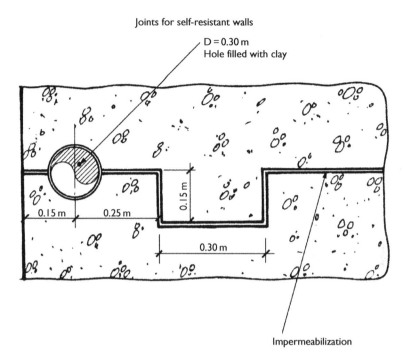

Joints for self-resistant walls

D = 0.30 m
Hole filled with clay

0.15 m

0.15 m   0.25 m

0.30 m

Impermeabilization

Figure 4.14 Joint for Burro Flume.

Figure 4.15 Burro Flume (Orellana Canal, Spain).

Longitudinal joints

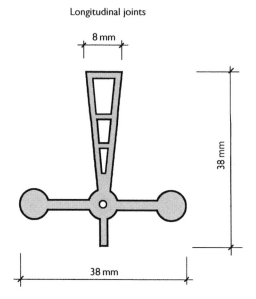

Figure 4.16 Strip for longitudinal joints.

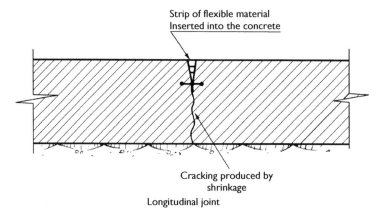

Figure 4.17 Longitudinal joint with a plastic strip.

4.3.3 Transverse expansion joints

Expansion joints have to meet different conditions to those of contraction joints and must be able to reduce their thickness by 1 cm.

The simplest form of expansion joint consists of a complete cut in the lining, with a minimum thickness of approximately 2 cm, which is then filled with suitable material (Figure 4.18).

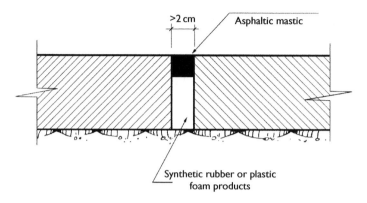

Figure 4.18 Simplest expansion joint.

The width of an expansion joint depends on the size of movements expected in the lining, together with the relative strains that the joint material is able to withstand. These may reach approximately 50% and will be examined in further detail later when describing the properties of these materials. In order to withstand movements in the order of 1 cm, the joint should have a minimum thickness of 2 cm.

It is not recommended that the full depth of the joint be filled with impermeabilization material. This is because, first of all, it is expensive since the joint depth is the same as the lining thickness of between 7 and 10 cm and, on occasions, may be even greater than this. In addition, the compression forces to which the joint is subjected due to thermal expansion of the concrete slabs cause some of the impermeabilization material to be forced out of the joint, which means poor joint operation, together with the entry of water and the solid particles it is carrying, which then compromise future movement of the concrete slabs.

To combat this effect, two different types of material are usually employed simultaneously. A material having a main property of being able to withstand large changes in volume, but without the need to provide great resistance to the compression forces is placed at the bottom of the joint. This material must be able to withstand all the previously described forces and loads.

The most suitable materials for the lower filling of the joint are bituminous felts and cork agglomerate; however, best of all are synthetic rubber and plastic foam products because they are not subject to decay problems.

Water pressure, acting on the outer joint impermeabilization element, will tend to force this material towards the bottom and press it against the lower filling in the joint.

These forces are transmitted to the lining's concrete walls by adherence. However, in principle, these hydraulic pressure forces are of little significance because, if we assume a canal having a depth of 3 m, with 2 cm wide expansion joints, together with a 3 cm thickness of impermeabilization material, the shear forces occurring between the joint bead and the two associated concrete slabs are given by:

$$0.3\,\mathrm{kg/cm^2} \times 2\,\mathrm{cm}/(2 \times 3\,\mathrm{cm}) = 0.1\,\mathrm{kg/cm^2}$$

As we shall see later, it is not difficult to obtain impermeabilization materials for the outer joint section that can withstand these shear forces under normal temperature conditions so that the actual joint itself is stable, without having to rest on the bottom filling, since this would not be able to resist due to its high strain coefficient.

The outer impermeabilization material may undergo strain due to creep, caused by the water pressure. This is not serious as long as no separation from the concrete occurs.

However, a serious situation does occur if the impermeabilization material peels away from the concrete, which often leads to joint failure.

In principle, the inner filling material does not possess any mechanical strength properties. Nevertheless, in a situation where the impermeabilization bead peels away and enters the joint, it would undoubtedly be of use in preventing total collapse and also facilitate repair work, which would simply require replacement of the impermeabilization material.

Among the classic type of expansion joints, which are not used for thin canal linings, but instead for thicker walls as in flumes and syphons, etc., mention should be made of the copper sheet type (Figure 4.19). The cutting and bending of the ends can be used to provide a certain amount of anchorage, which will prevent the sheet moving in relation to the concrete when

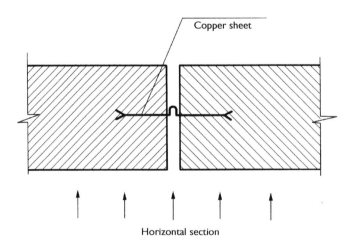

Figure 4.19 Copper sheet-type joint.

the joint opens. Bending the centre will make it possible for the sheet to adapt to this movement.

Because of its high cost and the difficulty involved in joining such different materials, this type of joint has been replaced by a similar one, which consists of strips of elastic, non-rotting material (Figure 4.20).

These joints produce good results, provided that the concrete between the wall and the impermeabilization material is thick enough to prevent it from breaking up. They are highly recommended for flumes etc. On the other hand however, they are not really suitable for thin canal linings and even

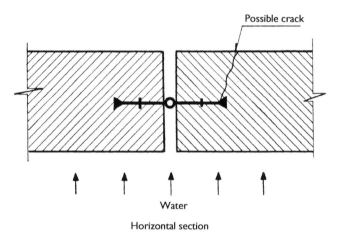

Figure 4.20 Joint with elastic strip.

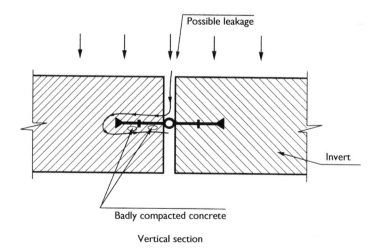

Figure 4.21 Problem of a strip horizontally positioned.

less so for bottoms. If the strip is horizontally positioned, the difficult concreting operation and the actual water that reflows while the concrete is still fresh produce a very poor bond between the flexible strip and the underlying concrete and this leads to a very easy path for leaks to take place (Figure 4.21).

4.4 Quality of materials required for joints

The severe conditions to which the materials employed in canal joints are subjected requires a very careful study of the properties they must possess.

In order to examine the impermeabilization materials for joints, we shall divide them into three groups.

The first group contains materials formed by mixtures that once they are applied to the joint, generally in the form of putty, have physical–chemical properties that are intermediate between those of the individual components before being applied. These are known as non-vulcanized materials.

The second group consists of materials made up of at least two basic elements, which are normally mixed just before application to the joint and which react together to form a completely different substance that is consistent, durable and adheres to concrete, and which possesses properties quite unlike those of the original components. This reaction is known as vulcanization, which is the transformation into a product that is somewhat similar to rubber and the materials are called vulcanizable.

The third group contain materials that are already prepared and form strips, which conserve their properties when fitted into the joints, together with their elasticity and strength, etc. The only operation they require is cutting to the necessary length.

Each of these three groups of materials is discussed below.

4.4.1 Non-vulcanizable putty for joint impermeabilization

Most of the products in this group are those that basically consist of asphaltic bitumen. A typical formula for these products is given by Van Asbeck (96). It contains 40% bitumen, with 40/50 penetration, 55% fine sand and 5% asbestos fibre with a length of between 2 and 4 mm.

The purpose of the asbestos fibres is to form a type of internal mesh that helps to prevent loosening of the material at high temperature. This phenomenon is perhaps the most dangerous in this type of material and it is therefore recommended that bitumen having less penetration be employed in hot climates.

It should be pointed out here that bitumen obtained by the fractional distillation of asphalt or petroleum is basically characterized by hardness, which is defined by the so-called "penetration test". This consists of a standard test in which a needle weighted with 100 gm is placed over a bitumen test piece at a temperature of 25 °C for 5 s (3).

At the end of this time, the needle penetration, measured in tenths of millimetres is used to define the bitumen.

It should be quite evident that the greater the penetration value, the softer the bitumen and vice versa.

Bitumen can be subjected to several classical processes, such as oxidation and fluidification, etc., to provide it with other properties, some of which enable it to be applied cold.

However, all the precautions that are taken to prevent loosening of the material with high temperatures tend to cause cracking at low temperatures and producing dangerous peeling away from the concrete. In order to avoid this, at least in part, it is essential that it is completely dry when applying the impermeabilization, which may even require the heating of the wet sides with a welding torch. The concrete can also be painted with bitumen derivatives that are sufficiently fluid to enable them penetrate all the surface pores and to establish a bond with the impermeabilization product that is applied later, usually hot.

The impermeabilization products just described do not normally comply with the conditions that are considered optimum for the impermeabilization of joints employed in canal construction (as far as temperature extremes are concerned). However, they have produced good results on most occasions and, in fact, are the most widely employed type throughout the world.

As with all materials, it is necessary for the engineer to carry out prior studies of the behaviour they are going to have onsite.

In order to accomplish this, two tests have been designed, which are intended to predict the behaviour of the impermeabilization material with regard to the two most serious dangers that threaten it. One is loosening caused by heat and the other is cracking and peeling away from the concrete due to low temperatures.

The first of the two tests is that of creep, which established the tendency of the material to loosen at high temperatures. In concept, the descriptions of this test provided by Van Asbeck (96) is similar to that of Juan Ortega (80) and which is included in the ASTM D1191 standard (2).

Basically, as described in the ASTM standard, this test consists of preparing a test piece, 6 cm long, 4 in width and 0.32 cm in depth using the corresponding bituminous material. This test piece is produced in a tinplate mould having the necessary dimensions. This test piece is positioned at an angle of 75° and held at 60 °C for 5 hours. The variation in length in centimetres is taken as the creep value for the test piece.

Our proposal more or less coincides with this description, with the test piece being prepared as shown in Figure 4.22, with suitable dimensions. It is then placed inside an oven at a constant temperature of 60°, which the American standard takes as being the maximum to which the lining will be exposed. It is evident that the design engineer should ensure that the inclination of the test piece coincides with that of the actual canal,

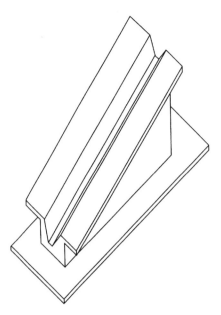

Figure 4.22 Piece for creep test.

together with its width, etc., so that the test reproduces the actual real case conditions as much as possible.

The other test, which is also accepted as being fundamental to the prediction of the behaviour of materials used in canal joints, is the one that establishes the level of adherence (2).

This is very closely linked to ductility and we believe that the most suitable method of performing this test is that described by Juan J Ortega (80), which coincides with the American ASTM D1191 standard. The test requires the preparing a test piece consisting of two mortar blocks, 2.5 cm × 5.1 cm × 7.6 cm (1 in. × 2 in. × 3 in.), separated by spacers to leave 2.5 cm × 5.1 cm × 5.1 cm, which is filled with the material to be tested (Figure 4.23).

Although not covered by the American standard, we believe it a good idea to previously paint the concrete with the bituminous product in fluid form if this is to be carried out at the work-site so that the test conditions reproduce the real situation as much as possible. Once the test piece has cooled, it is separated using a special mechanism that basically consists of jaws that hold the test piece and separate it at a very low speed. This mechanism is driven by a geared electric motor. According to the American standard, the speed should be 0.3 cm per hour during 4 hours at the various corresponding test temperatures. We are of the opinion that the temperature should be the lowest expected for the canal's location, which for most of Spain would be −5 °C.

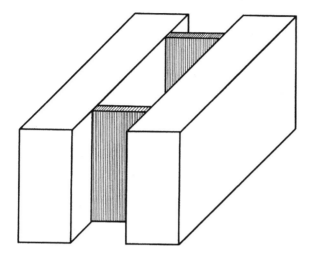

Figure 4.23 Adherence test.

It must be pointed out that the ASTM standard test speed may appear to be real for a canal with closed expansion joints; however, the true speed in the contraction joints would be much lower because the daily variation in width is much smaller.

According to Van Asbeck, the exact meaning of this test cannot be precisely defined in general. Nevertheless, it is quite evident that in our case it would clearly indicate whether the impermeabilization material would peel away from the concrete or not. The test should be performed several times in order to analyse the material's fatigue level.

Cracking may occur during the test on the outer faces of the impermeabilization test piece. Apart from being very significant in the prediction of the material's behaviour, there is no doubt that it can also be used to establish the minimum joint thickness and, above all, its depth, so that in spite of cracking appearing the level of impermeabilization continues to be effective. The dimensions adopted in the design for the joints should take into account the results of these tests.

4.4.2 Vulcanizable materials for filling joints in canals

Within this more or less ingenious classification that we are employing for the study of the various joint materials, this group includes all those that require mixing of two different components prior to application. The reaction between the two produces a new material (vulcanization), which then hardens and acquires properties that are quite different to those of the original components, including excellent impermeability power, deformability, resistance to temperatures and adherence to the concrete. One example of vulcanization often seen in everyday life is that of the epoxy resins that are used in the home to repair broken objects, in which two liquid components are mixed, and then solidify and harden in a notable fashion.

Many vulcanizable materials are hydrocarbon products, with more or less large molecules. When the two components are mixed, they react and the molecules bond together to form a spatial molecular network, which leads to changes in both exterior aspect and behaviour.

Many products consisting of rubber (synthetic or natural) and bitumen can be included in this group. Synthetic rubber generally obtained from petroleum can be perfectly combined with bitumen in the same way as natural rubber, improving its properties as a joint material. When powdered rubber is added to hot asphaltic bitumen, a fundamental change occurs in the bitumen's physical properties, with a reduction in high temperature penetration. This, quite evidently, makes it an excellent material for use in joints.

What is really desirable is a material possessing the same penetration at both high and low temperatures.

In view of these varied situations, the so-called "susceptibility factor" has been chosen in order to assess whether a material produces good or bad impermeabilization, and this is defined by the following expression:

$$S = \frac{\text{penetration at } 25\,^{\circ}\text{C with } 100\,\text{g, weight for } 5\,\text{s}}{\text{penetration at } 0\,^{\circ}\text{C with } 200\,\text{g, weight for } 60\,\text{s}}$$

This formula uses more favourable results with higher temperatures than it does with lower temperatures. What actually happens is that, even with good materials, the results are so different that in order not to handle susceptibility factors that are very high the test needle is loaded with different weights for different time durations.

It has been noted that the addition of rubber percentages of up to 5% can enormously improve the susceptibility factor, as shown in the following tables:

Asphaltic bitumen, ordinary distillation

Source	% Rubber	Susceptibility factor
Arkansas	0	4.55
	3	3.60
	5	3.30
Venezuela	0	3.80
	3	2.80
	5	2.48
California	0	5.72
	3	4.70
	5	4.06

Asphaltic bitumen catalytically oxidized

Source	% Rubber	Susceptibility factor
Wyoming	0	1.65
	1	1.38
	3	1.48
	5	1.50
Kansas	0	2.08
	3	1.57

It can be seen that the quality of the material as a joint component improves with the addition of small quantities of rubber, which produces the vulcanization effect, but above all, the use of the so-called "blown" or oxidized bitumen, which is produced by blowing air or oxygen through the hot bitumen, is highly beneficial.

Further information is provided in the table given below which compares properties:

Properties	Pure blown bitumen	95% Bitumen and 5% powered rubber
Penetration at 25 °C	37	29
Softening point °C	86	116
Penetration index	+4.4	+6.8
Viscosity at 150 °C CS	2300	50 800
Viscosity at 175 °C CS	510	11 000

Khasin (59) describes a series of comparative tests between epoxy resin based and vulcanizing products, mainly rubber- and bitumen-based, with the tests being carried out taking into consideration the effects of water, the effect of alternative wetting and drying and UV light.

With respect to all these loads, the epoxy resin-derived products provided the most constant results, whereas other vulcanizing products showed wide variation in their properties over time, which, in most cases, worsened.

The epoxy resin-derived product properties were maintained almost constant with a breaking stress of approximately $8 \, kg/cm^2$ and adherence strength to the concrete in the order of $2 \, kg/cm^2$.

The breaking stress of the rubber-based self-vulcanizing product began at approximately $12 \, kg/cm^2$, but after one year had been reduced to only $2 \, kg/cm^2$. The adherence strength to the concrete for these products started out around $2 \, kg/cm^2$, but fell to approximately $1 \, kg/cm^2$ after one year.

On the other hand, the epoxy resin-based products maintained their adherence to concrete constant even one year later.

The elongation of the epoxy resin-based products remained constant at 150%, whereas the rubber-based products improved with time and increased to over 200%.

Testing revealed that ultraviolet radiation was very harmful with respect to relative elongation. In the case of epoxy resin-based products, which commenced with 120% elongation, they maintained an elongation possibility of merely 40% after only a couple of months had passed.

It would appear that overall, the epoxy resin-based products provide the best levels of quality and that the weather effects to which the joint materials are subjected are very detrimental.

Khasin (59) studied the following product:

Epoxy resin: 100 parts by weight
Thiokol liquid: 300
Coal tar: 100
Carbon: 50
Polyethylene Polyamine: 10
Diphenyl guanidine: 1.

The properties of this impermeabilization product were as follows:

Breaking stress in kg/cm^2: 12–15
Relative breaking elongation in %: 100–120
Residual elongation in %: 4–5
Separation strength by adherence to concrete in kg/cm^2: 16–20
Ice resistance: $-40\,°C$
Heat resistance: $+80\,°C$.

However, it is our opinion that these specifications should not be taken into account in canal joint design because they are excessively favourable.

Moreover, because of the undoubted rapid ageing process undergone by all impermeabilization products, it is not recommended that any product be required to withstand strains in excess of its original dimension, even though laboratory testing prior to application would seem to indicate much greater possibilities.

This must be taken into consideration when establishing the joint's dimensions.

For the polysulphide types of self-vulcanizing products, which have been widely employed, the following specifications are normally used by both the Bureau of Reclamation and the state of California:

Work loads for an extension of 50%: between 0.6 and $1.3\,kg/cm^2$
For an extension of 100%: between 0.9 and $1.7\,kg/cm^2$

For an extension of 150%: between 1 and 2 kg/cm^2
For an extension of 200%: between 1.1 and 2.3 kg/cm^2.

These specifications apply to products with a minimum sulphide content of 20%.

It must be remembered that joint impermeabilization products currently have highly specialized formulas and that those from different manufacturers are not fully comparable. This is why it is very important to carry out comparison tests between the various products.

Polysulphide-based materials usually give good results.

4.4.3 Impermeabilization strips

This group contains all those elements that come in various forms of elastic strips, which are capable of withstanding significant elongation and contraction.

The materials indicated in Figures 4.18 and 4.21 belong to this type.

Deformation of these materials is produced on many occasions, not only because of the materials' own intrinsic properties, but because of the shape of the strip.

Normally these elements are made from either natural or synthetic rubber or plastic products, frequently PVC.

The state of California requires that the products used to manufacture impermeabilization strips have a tensile strength of the order of 140 kg/cm^2, together with a minimum breaking elongation of 450%.

The Bureau of Reclamation requires that elastic joint impermeabilization strips made from PVC have a minimum tensile strength of 120 kg/cm^2, together with a minimum breaking elongation of 350%. These were the conditions contained in the Specifications for the San Luis Canal, although on other occasions this same organization has specified a minimum elongation of 140 kg/cm^2. Great care should be taken to prevent some of these elements from coming into contact with other putty types of impermeabilization products because they will react in an unfavourable manner and may decompose.

It should be taken into account that the specifications for the various synthetic products can vary a great deal depending on their shape, thickness, dimensions and, above all, according to patents of manufacturing methods.

As an example, specification data for various plastic elements, provided by Airepetayan, are given below.

PVC membrane – Breaking load: 100 kg/cm^2. Breaking elongation: 180%

High-pressure polyethylene – Breaking load: 125–140 kg/cm^2. Breaking elongation: 600%

Polyisobutylene – Breaking load: $15 \, \text{kg/cm}^2$. Breaking elongation: 1000%

Polypropylene – Breaking load: $250 \, \text{kg/cm}^2$. Breaking elongation: 300–400%

Polyamide membrane – Breaking load: $300 \, \text{kg/cm}^2$. Breaking elongation: 70%

Opanol – Breaking load: $60 \, \text{kg/cm}^2$. Breaking elongation: 300%.

4.4.4 Materials for filling dead spaces in joints

The purpose of the materials examined in this section is to fill those parts of expansion joints that do not contain the actual impermeabilization product. From this it can be seen that the basic specifications do not refer to any impermeabilization properties, but instead to deformability due to compression caused by small loads and the ability to recover the original dimensions when the compression force is removed.

According to Juan J Ortega (80), suitable materials for this purpose, in the case of roadways, are cork, sponge rubber, mixtures of these two and bituminous fibres.

Other materials are also being employed for canals, such as PVC sponges, spongy plastic and neoprene, etc.

It is usually required that these materials are capable of reducing their thickness by 50%, with compression loads of less than $50 \, \text{kg/cm}^2$, which means that the expansion joints have a thickness that is double the expected movement at the joint.

The conditions to be met by these materials can be found in the ASTM D544-49 standard.

Among regularly employed bitumen-based compounds are those consisting of fibrous structures, high quality bitumen and where necessary, filler materials.

According to Juan J Ortega (80) the amount of bitumen in the putty should be a minimum of 70% by weight, or when the joint width is greater than 0.7 cm, the amount should be a minimum of 65%.

The putty is usually between two sheets of bitumen-impregnated felt or, in other cases, directly bonded to the fibrous structure forming a single element.

The amounts of these synthetic products correspond to those described in previous sections.

Chapter 5

Construction of concrete linings

5.1 Manual lining compaction and surfacing smoothing system

The simplest way to accomplish this, which is the one that is first thought of when concreting a canal lining, is to directly apply the concrete paste over the excavation of the canal box or transverse section and then compact it with rams and smooth it with a trowel before it hardens.

This system should be totally rejected because if the concrete is very wet, it will slide down towards the bottom of the canal slope, and if it is very dry, it will not compact well, leading to insufficient densities and even leaving significant cavities within. This lining will be permeable and have very little resistance against adverse weather conditions.

In addition to this, it is very difficult for the outer surface of the lining to be sufficiently flat and smooth due to the lack of a thickness reference for each point and that the thickness is adequate over the entire area.

To eliminate this last problem, one system that has been very frequently used consists of fitting wood ribs or masters, which adapt perfectly to the excavated canal section, with a thickness that is equal to the required lining thickness.

The fresh concrete is spread and smoothed with a plank supported on two wood masters, which are called so precisely because they direct the operation being carried out. The thickness is therefore that required, provided that the exactness of the ground excavation has been carried out with sufficient accuracy.

This same plank is then employed to strike the fresh concrete mass in order to compact it. However, the lack of plank weight and the very nature of the operation means that the concrete suffers from the same lack of compaction as in the previous case (Figure 5.1).

However, this system provides one significant improvement. If the section between two consecutive ribs is a slab, then the concreting is carried in alternative slabs; in other words, one is concreted and the next one is not. Once the concreting of two alternative slabs has been carried out, leaving the intermediate without any concrete, and the concrete has set and

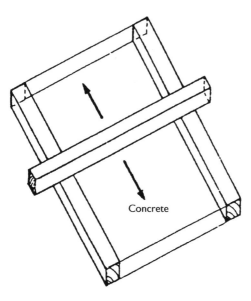

Figure 5.1 Spreading the concrete with a plank.

hardened, the ribs are removed and the empty area is then concreted. The only difference is that the smoothing and compacting plank is supported on the two neighbouring hardened slab surfaces.

It should be pointed out that this system allows the construction of canals with any shape of transverse section, without it necessarily having to be trapezoidal. The only requirement is that the excavation and ribs have the necessary shape.

The advantage turns out to be that without realizing it we have produced a contraction joint, which is precisely the joint between two consecutive slabs. It is a contraction joint because it permits a reduction of slab length due to cooling or shrinkage, but not their dilation because they are in contact with each other. The joint is impermeabilized simply by painting the concrete surface that was in contact with the rib with a suitable material, whether bituminous or not, chosen from those already examined, before concreting the intermediate slab.

From what has been said, it should be clear that the separation between the ribs should be equal to the separation required between the contraction joints in accordance with the regulations described in a previous chapter.

Previously, the use of a similar method was also very frequent, and this consisted of constructing concrete ribs buried in the ground, also in the direction of maximum angle of slope, and separated by the same distance that was required between the contraction joints. First of all, these ribs were used to exactly define the profile of the excavation under the lining since it was only necessary to stretch a length of cord over two consecutive ribs in order to immediately establish whether further excavation was required or not.

For this reason, maximum attention was paid to constructing these ribs at their exact elevation since they were to define the final support base for the lining, which again was the reason why they were called "masters". It was only then necessary to ensure that the canal lining thickness was sufficient to establish the canal lining surface at its correct elevation. All that was required was a plank employed as a mould or formwork set on top of the concrete ribs in accordance with the maximum canal slope angle, which would then laterally contain the concrete for the first series of slabs.

This system concretes the lining by alternative slabs and this turns out to be an advantage because the initial opening of the contraction joints is less since the previous slabs had already set and undergone shrinkage before the intermediate slabs are constructed.

In addition to the poor degree of concrete compaction obtained with these manual compaction systems, proof of which is sometimes given by the fact that plants sometimes grow in the lining cavities, there is also the problem in which the constructor gives way to the temptation of extending a layer of mortar over the lining in a form of roughing-in intended to cover up these cavities. However, this roughing-in operation does not bond sufficiently to

Figure 5.2 Alberche Canal (Spain).

the underlying concrete and tends to loosen with temperature changes, and its use should be totally outlawed. These defects can be seen in Figure 5.2, which shows the Alberche Canal in Spain.

Because of the poor quality it produces, manual compaction of concrete should always be prohibited and only concrete that has been vibrated at speeds of 5000 vibrations per minute or higher (which leads to problems that we shall examine later) or vibro-compacted should be accepted.

5.2 Quick assembly and disassembly formwork systems

The vibration of concrete used for canal linings involves either horizontal surfaces or those with very little inclination (so that there is no danger of the material sliding downwards), or it is necessary to employ formwork (in other words, moulds), which contains the concrete, while it is being vibrated and subsequently sets.

The problem with the use of formwork for canal lining construction is the very high percentage of the overall cost involved.

If we consider, for example, that the lining is 10 cm thick, the formwork surface per cubic metre of concrete is $10\,m^2/m^3$ (since the formwork is only required for the upper surface as the lining is supported on the ground), which is a comparable quantity to the concrete pieces with most formwork that are employed in building construction, but in general much less than that used in hydraulic works (dams, other large volumes of concrete and walls, etc.).

This means that any reduction in cost of formwork per square metre will have a tremendous effect on the final cost of the canal.

In order to achieve economic formwork prices, the first important thing is to be able to use it many times over, taking advantage of the fact that there is a great length of canal all with the same or very similar characteristics, and so it should be constructed from durable material, such as metal sheet. However, the most important point is to produce a very fast assembly and disassembly system in order to save on the labour costs involved in these operations. Moreover, it is also necessary for the time that a given piece of formwork is in position, at a given point, be as short as possible so that it can then be used at another point. This will reduce the required number of pieces and the amortization of the capital invested in its purchase will be reduced.

Figure 5.3 shows a system that was successfully used with the Orellana Canal, Section C, which consists of a series of metal beams anchored to the ground and located according to the canal's maximum slope angle. These are intended to support the flat panels (Figure 5.4) of the formwork, and have hooks that are fastened to the beams, and which are very easy to install and remove.

Figure 5.3 Steel beams to support the formwork.

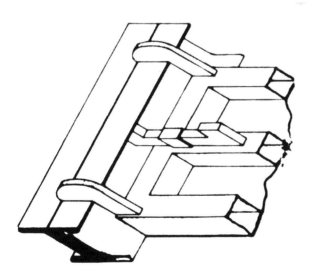

Figure 5.4 Panel of formwork.

A free space is left between the formwork surface and the ground, which has a thickness that is equal to that of the lining that is to be constructed and which is filled with concrete. The beams must be anchored to the ground so that they do not lift due to the hydrostatic thrust produced by the fresh concrete, which is increased by the vibration operation.

The vibrators used for this operation are the poker type, which are inserted into the free space between the formwork and the ground. This means that this free space, in other words, the final lining thickness, must be sufficiently large so that the vibrator fits inside, with an adequate value being 12 cm, a value that has to be taken into account during the canal design phase.

The time during which a particular piece of formwork remains in place at a specific point is strictly that required for the concrete below to harden sufficiently so that it does not slide and also to prevent any possible damage when the formwork is removed.

The following slab also requires the formwork support beams and this means that when the beams are removed, a space remains, in which the transverse joints may be constructed.

5.3 Transverse sliding formwork systems

Since it is required that the time the formwork remains in place is that which is strictly necessary for the concrete to harden sufficiently so that it does not fall to the bottom, and at the same time leads to an economic number of formwork units, it is possible to employ sliding formwork systems, which are moveable, with a very slow, yet sufficient, displacement speed of, for example, 2 m per hour (the actual value should be determined by trials).

The simplest method of constructing a canal lining using this procedure is that described by the International Commission for Irrigation and Drainage.

This method consists of placing wooden ribs on the ground in the direction of maximum slope angle, with a thickness equal to the thickness of the lining to be constructed. A wooden or metal tray is supported on these, which can then be displaced upwards under the power of a winch located on the canal's upper bank. This forms the sliding formwork, which leaves a space between itself and the ground underneath to be filled with concrete (Figure 5.5).

As it is poured, the concrete should be vibrated with a poker vibrator, which means that lining thickness must be large enough to accommodate it (for example, 12 cm as a minimum).

The wooden tray is ballasted with sandbags to prevent it rising due to the hydrostatic thrust of the soft concrete.

The tray ascent speed should be slow enough to allow the concrete being uncovered to have hardened sufficiently so that it does not flow downwards. The exact speed will depend on, among other factors, the amount of water

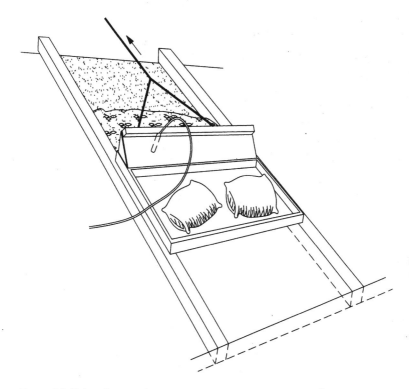

Figure 5.5 Sliding formwork.

in the paste and should be established for each case by carrying out a series of trials.

When the concreting process for a given slab has been completed, the ribs are removed and the operation is continued employing the alternative slab system. When this stage has been completed, the intermediate slabs are concreted using the same process, with the only modification being that the trays are not supported on the ribs, but instead, on the adjacent, already hardened and finished slabs (Figure 5.6).

Before concreting the intermediate slabs, the surfaces of the first slabs should be painted with an impermeabilization product in the zone that is perpendicular to the ground and which make contact with the next slab, forming a contraction joint. The separation between the ribs should therefore be that required between the contraction joints.

Because they are vibrated, this system produces excellent quality lining concrete, with minimum means, since only a concrete mixer, vibrator, winch, beams and a wooden tray are required. The accuracy of the final

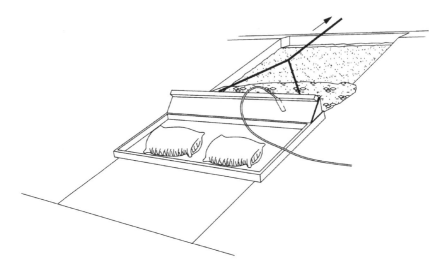

Figure 5.6 Concreting intermediate slabs with a sliding formwork.

Figure 5.7 Machine with sliding formwork.

lining elevations depends on the rib placement, which is also used to verify whether the excavation profile is correct or not.

Because of its simplicity the description just given is that of a system that can be employed on non-developed land. However, more perfected machines have been produced, based on exactly the same idea, which provide improved performance. All of them employ metal sliding formwork which is fitted with powerful vibrators, so that the lining thickness may be less than that stated earlier. Figure 5.7 shows one of these machines.

5.4 Longitudinal sliding formwork systems

There is another canal lining concreting system employing sliding formwork which is very similar to the one described above and also based on it. The only real difference is that the formwork is displaced in a longitudinal direction to the canal instead of transversal.

The formwork can cover the entire canal section or just one slope.

In the latter case, one slope is concreted first and then the other is completed using the same machine. The bottom is concreted last, although there is no reason why it cannot be constructed first. In both cases, great care must be taken with the impermeabilization of the joints between them.

The formwork in this case may be displaced along two railway track–type rails, where one is located on the canal's upper bank and the other on the bottom. Perfect alignment and levelling of both is essential since the accuracy of the canal termination depends on this.

The flat formwork can be fitted with powerful surface vibrators, so that the lining thickness no longer depends on the need to have sufficient space for inserting the poker vibrator.

Figure 5.8 shows the machine used for concreting one of the Bajo Ródano canals in France. The rails weigh 45 kg and are supported on short sleepers measuring $50 \, cm \times 20 \, cm \times 15 \, cm$. The machine's displacement speed was around 30–50 cm/min and had a total input power for the operations to be carried out (displacement, concrete distribution and spreading, together with vibration) of 125 hp, of which 40 hp corresponded to the displacement motors.

The actual machine itself requires seven men:

1 driver
1 concrete distribution carriage operator
1 concrete output control operator
2 longitudinal joint operators
1 mason for termination control and finishing
1 assistant to clean up any concrete waste.

Figure 5.8 Longitudinal sliding machine for canal construction.

The joints were constructed using a disc saw when setting had commenced, as shown in Figure 5.9.

When this machine is employed for concreting the slopes, the bottom is constructed by traditional means and, of course, the concrete must be vibrated.

Longitudinal sliding formwork machines do not always move along railway track rails, they often run on caterpillar tracks as do tanks.

However, it is also possible to employ the formwork to construct the entire canal cross section by employing machines, such as the Rhaco that was used to construct the Navarra Canal and which is shown in Figure 5.10.

Under the concrete lining is located an impervious membrane such as those mentioned in Chapter 8.

It has the tremendous advantage that the longitudinal joints between the slopes and the bottom of the previous method, which are always delicate points that have to be carefully impermeabilized, are no longer required.

Construction performance is improved since this affects the entire section.

As a counterpoint, there is the problem of the lining machine being adapted to the varying canal shapes and sizes because the formwork dimensions cannot be easily changed. It has to be remembered that the machine consists of the complete formwork section, conveyor belts for the fresh concrete, together with distribution and compaction elements, etc.

Figure 5.9 Longitudinal joints constructed with a disc saw.

Figure 5.10 Construction of the Navarra Canal (Spain).

However, there are canal concreting machines that incorporate the possibility of modifying the slope angle by means of joints between the base and sides, which are then strengthened. The bottom width can also be varied to a certain extent by adding or removing bolted-on sections.

In spite of this, canal lining machines normally have very little flexibility with respect to being coupled to differing sections, so that they are more recommended for very long canals that have a practically constant section (in the case of a water transfer canal, such as the Tajo-Segura), than for irrigation canals which have a telescopic variation of sections in order to adjust to the transport of ever-decreasing flow rates.

When there is little variation in flow rate, it is possible to resolve the problem of varying canal sections with one of these machines based on constructing the lowest capacity stretch with a section of the same base width and slopes, but with less depth. This will mean that the formwork and the machine in general remain the same, but the machine operates in a somewhat higher position. Of course, this will require that type of machine to be employed is established during the drawing-up of the project.

All these machines are fitted with a round visor at the front that plays the role of formwork and as it advances forces the concrete extended in front to spread underneath between the machine itself and the ground.

Most of these machines employ poker vibrators in front of the visor that fluidify the fresh concrete and the compaction effect is produced by their own heavy weight.

Since the energy of this type of compaction is distributed over the entire layer thickness, it is recommended that this be thin. In this way, some fifty years ago thin concrete linings appeared on the scene of approximately 8 cm, which were frequently used for both types of machines. It must not be forgotten that these thicknesses require much smaller aggregate that often requires the stone to be broken up, so that the savings indicated by the reduced volume of concrete is only apparent.

The thicknesses are, of course, sufficient to guarantee the required impermeability for the canals with the indicated considerations with respect to contraction joints, provided all the necessary precautions are taken to reduce the danger of cracking. On the other hand, the concrete is incapable of withstanding the bending produced by differential settling in the ground, independently of the lining thickness. This pernicious effect must be combated by excellent quality slope compaction, together with the clearing of the vegetative layer or poor quality excavations. It must not be forgotten that the level of compaction has to reach 95% of the Proctor value obtained in the laboratory (Chapter 3).

In any case, the compaction obtained only by the machine's own weight or by a combination of its own weight and poker vibration may not be enough to achieve the desired concrete and so other additional measures must be adopted to accomplish this. Among these is that of using concrete mixtures with a higher percentage of cement than normal (for example, a cement dosage of 300 kg per cubic metre of concrete is quite frequent with this type of machine, when only 250 kg or even less are sufficient for linings executed employing other procedures). The amount of sand is also sometimes increased and, in any case, air entraining agents are used to provide a paste with 4 or 5% occluded air. The purpose of all these measures is to increase the greasiness of the paste, increasing its workability because the very small grains of cement and sand, together with the air, will efficiently work in the form of ball bearings to facilitate internal movement and the dispersion of the gravel to achieve a more compact mixture.

These machines normally work with concretes having a cone descent of 5 or 6 cm.

Good concrete strengths are obtained with this type of machines, attaining values in excess of 200 kg/cm^2 at 28 days.

When canals are constructed with such thin linings, any small error in excavation accuracy of only 2 or 3 cm can lead to a very significant percentage variation of the volume of concrete employed (for example, 30%), which is quite unacceptable in view of the expensive nature of the unit of concrete volume.

This is why very strict measures should be taken to ensure the excavation accuracy and that of the upper elevations of the lining surface.

The lining or paver machine is always preceded by a trimmer, the purpose of which is to trim the excavation carried out by bulldozers or excavators, which perform a first rough excavation, leaving an excess of some 20–40 cm, which provides temporary protection against the weather for the underlying soil. The trimmer may be designed to excavate one slope first and the other afterwards (Figure 5.11) or to excavate the whole canal section (Figure 5.12).

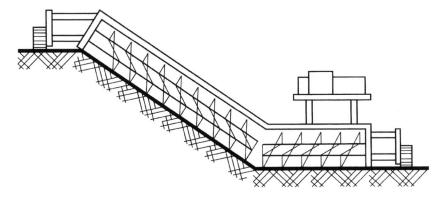

Figure 5.11 Trimming machine sketch running on caterpillar.

Figure 5.12 Trimming machine excavating the whole cross section of the Tajo-Segura Canal (Contractor: Ferrovial S.A.).

The distance left between the trimmer and the paver should not be too large in order to prevent the accurately excavated ground from being negatively affected by weather, but neither should it be too small because any momentary halt of the trimmer, for whatever reason, would also cause the paver to stop working. A minimum distance of some 200 m is usually considered reasonable, although this, of course, depends on the machines' working speed. If this distance is measured in machine advance time, there should not be more than 12–24 hours between the two, but can be less if the weather conditions are not good.

The trimmers usually perform their work by means of worm screws that operate in parallel to the maximum slope angle, which shave off the excess ground as they rotate. At the same time, a bucket chain collects the excavated soil and tips it onto conveyor belts, which then dump it onto lorries or onto tips located in suitable areas. All this equipment is installed on a metal structure that has the shape of the canal's transverse section.

Since the linings are very thin, it is very important for the excavation profile to be very accurate and on the outer lining elevations. This is achieved by means of an electrical device that drives the various motors faster or slower.

The concreting machines that work on the complete canal section normally run on pneumatic tyres or caterpillar tracks, only very rarely do they operate on railway track rails.

The two side surfaces of the ground, along which the wheels or tracks run, are not levelled, since this would involve significant extra costs. The adjustment of the trimmer and paver machines to the correct levels is automatically obtained through the suspension to each wheel or track, which lengthens or shortens using a hydraulic piston mechanism.

The same thing happens with respect to the forward motion of the machines, which cannot be allowed to vary at all from the established line of travel. This is achieved by the employing different engines to drive the left wheels or tracks and the right wheels or tracks. This means that a greater or lesser relative speed on one side with respect to the others would cause a modification to the machines' movement.

In order to obtain automatic adjustment of the direction and levelling of these machines, all that is required is a controlling string line along the sides of the future canal, consisting of a strong nylon line fastened to stakes driven vertically into the ground separated by distance of 5 m for straight sections and 3 m for curves. This line must be perfectly levelled and set parallel to the canal bottom axis.

The machine is fitted with a sensor consisting of a curved piece that rubs along the controlling string line. Depending on whether the line comes into contact with the left, right, top or bottom of the sensor, it transmits a different signal to the engines that adjust the direction of travel and the elevation. These react to the signals by adjusting their values so that the canal remains at the desired elevations and alignments.

There are several different brands that manufacture machines, such as those described, for constructing entire canal sections at one time. The average forward speeds normally lie between 15 and 30 linear metres per hour, which can be doubled at peak production times.

It can be understood that within a relatively short space of time, it is possible to construct a fairly long stretch of canal with a single transverse section. The only remaining problem, as has already been stated, is that resolving changes in the size of the section, since these machines are mainly designed for operation in very long canals with very few modifications to their section.

Another very important problem to be resolved in the concreting of a canal lining is that of concrete curing. If it is important for any concrete to retain sufficient moisture to produce the chemical setting process, it is even more so in this case because the reduced thickness aggravates the problem and encourages rapid drying of the entire mass. The lining lies directly on the ground and this will lead to further water loss due to absorption.

A good level of concrete lining curing is obtained either by continuous watering of the surface using perforated hoses, for example, or by preventing water evaporation by means of an immediate application of a special curing paint which is similar to enamel and impedes the passage of water vapour.

This paint is usually whitish and this enables painted areas to be easily distinguished from unpainted ones, which will facilitate inspection work. Figure 5.13 shows this curing paint being applied.

The problem of reducing the water losses due to ground absorption still remains, however. A very thin sheet of polyethylene has been laid underneath the lining in some canal construction works. We do not, however, agree with this method since it encourages downward sliding of the lining, and the loss of the favourable connection with the ground. We would prefer to employ very careful ground watering immediately prior to the concreting operation, taking care to avoid runoff gullies. It is also necessary to resolve

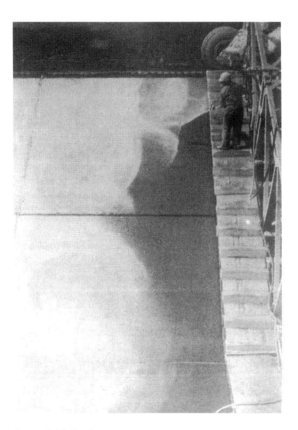

Figure 5.13 Applying curing paint.

the matter of joint construction, both for machines that produce a complete canal section and those that line only one slope at a time. Different procedures are followed for transverse contraction joints as for longitudinal ones.

Transverse contraction joints are produced by cutting the fresh concrete using a vibrating blade fitted to the machine at regular intervals, with a depth of approximately one-third of the lining thickness. An impermeabilization product is applied once the concrete has set.

The longitudinal joints are made either by leaving a longitudinal cut to a depth of one-third of the concrete thickness as the machine moves forward, or by the system described in the section on joints, consisting of leaving a suitable strip of plastic material within the fresh concrete as it is extended, which will weaken the lining and localize the cracking at this point, and also producing sufficient impermeabilization because of its shape, which provides good anchorage to the concrete.

The expansion joints are constructed independently of adjoining operations when a given stretch is completed.

5.5 Concreting systems for circular canals

As we shall see in Chapter 10, canals with a circular transverse section possess certain advantages, such as being more economic since they have less wet perimeter for the passage of water for a given useful area and hence, a reduced lining volume.

But in addition to this, they possess greater stability from the ground point of view, since if a flat slope slides it will tend to adopt a circular shape, known as a "Swedish Circle", which is similar to that which we gave to the canal at the beginning. This means that from the very beginning we design the canal with a transverse section that is quite similar to the optimum form from a stability point of view.

Also with respect to the lining, it has greater strength because the circular shape better approaches the line of the ground thrust loads than the trapezoid section. This is due to a phenomenon similar to the fact that an arched bridge is better prepared to withstand loads than one with a straight beam.

One might suppose that the construction of a circular section would be more difficult, but, in fact, it is not.

With regard to excavation operations, it has to be pointed out that backhoes have blade movement that adapts very well to circular sections so that an experienced operator can produce a circular section more easily than a flat one.

The lining concreting work is no more difficult either.

First of all, it is possible to make use of concreting machines, such as those already described, that employ sliding formwork for the entire canal section, but with a circular shape instead of trapezoid.

Secondly however, it is possible to employ purpose-built machines for this type of work, such as that designed by the engineer Barragán.

This consists of a metal portico that moves in parallel to the canal over metal rails, which are located on each side of the canal excavation (Figure 5.14).

A metal arm hangs from the portico lintel beam, and which is able to rotate around its connection shaft. When it rotates, its end describes the circular surface for the transverse section of the canal we wish to construct.

A metal tray is fixed to the end of the arm, and this forms the sliding formwork. It has a somewhat curved visor that pushes against the fresh concrete laid on the canal bottom, which is then pushed between the ground and the formwork when the arm rotates in an upward direction.

Powerful surface vibrators are installed on the outer section of the sliding formwork and these are able to produce such an energetic level of vibration that they can compact thicknesses that are much greater than those obtained using the previously described machines (for example, 25 cm) (Figure 5.15).

This allows for (it is also recommended) the use of very dry concretes (with nil settling and water–cement ratios in the order of 0.45), which are the ones that attain the highest strengths with these vibration powers. At the same time they are very compact and only slightly permeable and which are well maintained without any sliding, even on relatively steep excavations.

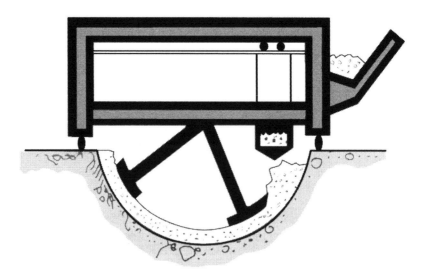

Figure 5.14 Machine for circular canals.

Figure 5.15 Sliding formwork for circular canals.

Generally, instead of just one arm, there are two symmetrical ones, each of which sweeps half the section from the lowest straight line upwards.

These machines also require an accurate prior profiling of the excavation in order to ensure that there are only very small differences in the lining thickness.

This is achieved using another trimmer, which is very similar to the paver. It also consists of a portico mounted on the same rails and also is fitted with two arms on the lintel. The difference here lies in the fact that fitted to their ends, instead of formwork, there is a wheel driven by an independent engine. A set of blades are mounted around the periphery of this wheel, which "shave" the excess soil away from the rough-levelled ground previously left by the backhoes. The rotation of the arm fitted to the portico allows the entire section to be covered in order to produce a perfect circle (Figure 5.16).

It should be quite obvious that the accuracy in the final canal elevations and alignments depends almost exclusively on that of the direction and levelling of the rails; to this end it is essential to have the permanent services of a topographic team to assist in their installation and to continually carry out the relevant inspections.

Figure 5.16 Barragán rimming machine.

The concreting of this type of canal is therefore carried out in accordance with transverse bands having the same width as the formwork tray. This dimension should coincide with the necessary separation between the transverse contraction joints, or slightly less and so allowing the machine to construct the joint between two bands. However, the actual width is generally approximately half the distance between the joints in order to make the vibrating tray more manageable. We do not see any inconvenience in doing it in this way, provided that the contraction joints are constructed in the contact zones between two contiguous bands.

We believe to be extremely wrong to construct the bands one after the other without these contraction joints. One can be quite certain that once the setting shrinkage has taken place and the temperatures drop, the concrete will contract and the lining will open at its weakest points, which are precisely the joints between the construction bands. This will lead to significant filtration, a fact that has been proven in practice.

The design of this type of circular section should be made taking care to select the upper angle forming the circular arc with the ground in a fashion that guarantees stability, for example, by making it equal to the internal friction angle.

In Figure 5.17 can be seen the almost completed Zújar Canal, which was constructed using the Barragán machine.

Figure 5.17 Zújar Canal (Spain).

5.6 Lining thickness

In canals, and probable more so than in any other civil works, the ground, project and its execution must be in perfect harmony.

Before deciding on the lining thickness to be adopted, it has to be made quite clear that this is never established by the requirement to obtain greater or lesser impermeability. Any practical thickness will be sufficient to produce the desired level of impermeability, provided the lining does not crack. For this reason great care must be taken with the various joints, which are the very elements that prevent this phenomenon from occurring.

The necessary considerations are basically economic and the most important ones are the following:

In order to spread and compact the concrete, it is necessary that the maximum aggregate size does not exceed 35% or an absolute maximum of 40% of the thickness. A smaller thickness does not always result in cost savings because reduced thickness values will require an aggregate size that could involve expensive crushing operations in special mills, or the elimination of a large amount of excessive sizes.

Moreover, it should be remembered that the contraction joint separation is proportionally dependent on the lining thickness so that if a very small lining thickness is adopted, the number of joints will increase, together with the final cost.

The type of concreting machinery to be employed is also very important and which significantly depends on whatever is available at the best price, together with its possibilities. We have also already explained that longitudinal sliding paving machines, which compact the concrete due mainly to their own weight, can only construct thin linings. This is the main reason why, several years ago, lining construction shifted from 15 to 20 cm of thickness to only 7 or 8 cm. On the other hand, in sliding formwork systems in which the compaction operation is performed by poker vibrators and which are inserted between the ground and the formwork, the lining thickness has to be enough to allow entry of the vibrator and is usually around 12 cm.

However, it has already been stated that the Barragán machine, with its circular sections, does not have any problem in constructing concrete sections having a thickness of 20 cm. It makes no sense to design greater thicknesses, expect in a situation where we move from normal linings to actual walls that are resistant to ground thrust.

It should not be forgotten to take into account the greater or lesser capability of the machine being able to adapt to canal sections of differing sizes and compare this characteristic with the lengths of each section to be constructed.

5.7 Concrete quality

Although it might appear that this point should have been covered earlier, it is included here because it is something of a result of everything that has been dealt with up to now.

The opinion of the Bureau of Reclamation, with regard to the extremely harsh conditions to which concrete is exposed to in canals, has already been made clear in Section 4.1. This obviously requires extreme caution to be adopted.

One of the most basic goals is to prevent cavities forming in the concrete as this will not only prevent filtration, but also stop plants rooting inside them and icy conditions from destroying them, etc. Moreover, it can be shown that when good levels of compaction are achieved with the consequent absence of cavities, not only impermeability, but also compression and tensile strength, together with durability, etc. are improved.

In order to achieve this, a grain size mixture must be employed that is as closed as possible so that there are almost no cavities. This occurs because a certain size of aggregate will fill the cavities produced by larger sizes, and the cavities they leave will be filled by still smaller aggregate sizes and so on. This will clearly require sufficient quantity of each size in order to fill the corresponding cavity size, but never more than this because it would cause certain sizes to separate larger ones, thus losing the lattice structure where some fit into others, which is essential for obtaining high strength.

Low water–cement ratios are also required for good levels of strength. This means ratios of around 0.48–0.55, depending on weather conditions and, above all, on the compaction power, either vibrated or rammed earth methods. We have already stated that the Barragán machine uses water–cement ratios of approximately 0.45, with nil settling. Settling values usually tolerated with normal sliding formwork machines lie between 4 and 8 cm and from 4 to 10 cm in structural concrete and walls.

Good levels of compaction are difficult with such dry concretes, but nonetheless essential and are obtained through powerful vibration employing poker vibrators that are inserted between the formwork and the ground. Surface vibrators installed on the formwork can also be used, although ramming of the fresh concrete is also used, provided it is a very thin layer, or a combination of vibration and ramming can be used.

In general, these water–cement ratios require the use of air entraining agents to facilitate compaction operations. The air bubbles that form act as miniature ball bearings, which assist the movement of some aggregates grains over others in order to fill in the cavities. Use of air entraining agents should not be overdone and the maximum usage values are provided below:

Maximum aggregate size (mm)	Air volume (%)
19	6
25	5
38	4.5

The occluded air percentage may have an error of ±1% in the first case and ±0.5% in the other two.

In the most thorough scientific analysis carried out on concrete, it was established that its strength is really a function of the *cement/(occluded air + remnant water)* and not just simply of the water–cement ratio, with values referring to the mixture that is commencing to set. When this ratio is decreased the concrete strength also decreases. The efficiency of the air entraining agents is that their use produces improved compaction because the mixture is more workable. This is very favourable with respect to strength and the secret to success lies in the use of air entraining agent dosages that produce greatest benefits because of the improved compaction, which compensates for the reduced strength due to a poor cement/(air + water) ratio so that the result is favourable to increased strength.

In addition, in order to facilitate internal aggregate rolling when using sliding compaction machines, the amounts of sand and cement are forced somewhat because to a certain extent, these also act as internal ball bearings. At this point, all that has been described concerning lining paver machines employing sliding formwork should be taken into account.

However, these are not the only methods of obtaining good concrete workability in order to produce good, simple compaction. Plastifying agents are also available. These are large molecule organic products with one hydrophobic end and one hydrophilic, which causes them to align themselves on the grain surfaces producing a lubricant effect, together with a dispersion of the cement grains and this results in a minimum quantity of water being required to wet them. This means that the water–cement ratio can be reduced, which will therefore increase the strength, but with the same paste workability. This reduction in water may reach between 5 and 15%.

The plastifying products are resinous soaps or byproducts from paper paste manufacturing.

These plastifying agents will somewhat delay setting times, which should be taken into account when deciding when to remove the formwork.

In general, the air entraining agents are more beneficial with low cement dosages (less than 250 kg/m^3), whereas the plastifying agents are more useful with high cement dosages.

It must not be forgotten that with respect to cement amounts, longitudinal sliding formwork systems frequently employ a dosage of 300 kg/m^3 of concrete. However, other systems usually employ rates of between 200 and 250 kg (250 kg in the case of circular canals constructed with the Barragán machine).

Some information about the types of cement that should be employed is given below.

The basic components of a clinker, which is a material produced by a mixture of clay and limestone at high temperatures and which is then ground down to produce Portland cement, are tricalcium silicate, bicalcium silicate, tricalcium aluminate and tetracalcium aluminoferrite.

The tricalcium silicate forms approximately 40–50% of an average type of clinker. This is the component that develops a high initial strength, but has a high hydration heat level, which could lead to cracking if the concrete mass is large. Since canal linings are normally very thin, except the lateral walls, heat evacuation is guaranteed and this problem may be safely ignored and in general for other canal works since they do not usually include any great thicknesses.

Those cements containing the described proportions of tricalcium silicate are very vulnerable to selenitic water attack and they should not be employed when the ground contains gypsum. In order to correct this problem, there are cements in which the tricalcium silicate is reduced to less than 5%, with a corresponding increase in the amount of monocalcium silicate.

The monocalcium silicate accounts for between 20 and 30% of a normal clinker. It is slower in attaining high strength levels, but on the other hand, develops very little hydration heat, which in general is not a significant point in canal construction. Its great advantage is that of chemical stability, for which reason it is less affected by attacks due to the presence of gypsum.

The tricalcium aluminate forms some 10–15% of the clinker weight. It possesses a very high setting speed, a great deal of hydration heat and shrinkage and is also very susceptible to sulphate attack. The cement should be rejected if the stated proportion is exceeded and it is of little use in the presence of gypsum. There are, in fact, sulphate-resistant cements that are based precisely on the absence of tricalcium aluminate (for example, "Sulfadur" cement). It is a curious fact that to combat the very high speed setting rate, a little pulverized calcium sulphate is added to the clinker in order to reach a compromise situation.

The tetracalcium aluminoferrite is present in the clinker without the manufacturers paying much notice to it; however, it is necessary for the actual production process. It represents between 5 and 10%. Although it adds little to the overall mechanical strength, it has a low level of hydration heat and is able to withstand sulphate attack better than the other clinker components.

From what has been said up to this point, it may be assumed that the high setting heat and high shrinkage levels are as serious as they could be in other, much more voluminous hydraulic works (in the case of dams, for example). If we were to be very meticulous, we would have to recognize that the high levels of shrinkage could influence the thickness of the canal's transverse joints, which in fact is something that is not often taken into consideration. Finally, high levels of tricalcium silicate are not especially harmful for canal linings unless gypsum is present.

If there is any danger of gypsum being present, then cements containing only a little tricalcium silicate and correspondingly more bicalcium silicate should be employed, and also with very little tricalcium aluminate, knowing well that the strength will be attained much more slowly. When sliding machines are employed to execute the linings, the concrete spread by the machine must have sufficient strength and stability to prevent it from sliding down the slope. We feel that any delay in the production of the concrete's strength when it contains a lot of bicalcium silicate hardly has any affect on machine operation because the slopes are always gentle; in addition, because basically the stability of the compacted concrete is in itself much greater than fresh concrete and this is independent of the speed of strength increase. Nevertheless, it is recommended that checks be carried out to ensure the concreting speed is compatible with the paste hardening speed.

The main problem occurs when gypsum represents a serious danger to the concrete. In this situation, in addition to employing the described solution of using Portland cements in which part of the tricalcium silicate has been replaced by bicalcium silicate and the tricalcium aluminate has been eliminated, there are other cement types that may be used, among which are those to which other products have been added.

During cement setting, a chemical process that produces heat, the soft, wet concrete paste is transformed into a rock-like substance, producing free lime. This is precisely what reacts with the calcium sulphate or gypsum forming the so-called "Ettringite", which has the property of increasing the volume and destroying the concrete.

The products that are added to the cement in order to resist the effect of the gypsum generally act by combining with the free lime before it is able to form the Ettringite. The two most frequently employed products are basic slag from foundries and the pozzolans, which are crushed natural volcanic products. Both are silicates that react with the free lime forming calcium silicates.

Slag naturally contains some lime and can therefore set if it is wetted. However, in order to act efficiently, they also require free lime, such as that released by the cement on setting. They are employed in cement in highly variable proportions of between 20 and 80%.

All of them, depending on the amount of slag, have a very low level of hydration heat and little shrinkage. These are very significant properties for large-scale concreting operations (dams, for example), but are of very little interest in canal construction. Their main inconvenience is that, due to the very slow rate of setting, the linings must be watered or measures must be adopted to conserve the moisture during a relatively long period of time, for example two weeks; this would be a very serious inconvenience in canal construction. Overall, this type of cement cannot really be recommended for canal execution.

The other product that is added to cement because of its beneficial properties with respect to sulphate attack are the pozzolans.

These are natural minerals, with a volcanic origin, which contain silica and alumina. Another similar product is fly ash produced by the combustion of coal, either in lumps or crushed. If these are finely ground and mixed with the cement, on setting they will combine with the free lime and enable the concrete resist sulphate attack. The liberation of free lime is not an instantaneous process, but one that lasts a long time. As this lime combines with the pozzolans or fly ash, the resistance increases, together with the compactness, so that beneficial effect is incremented over time. The amount of pozzolans that may be added to the cement is very variable and can exceed 20%, with the hydration heat decreasing as the pozzolan percentage increases. However, the hardening process is much slower than that for normal cements, which obviously must be taken into account in canal construction because of the influence on the amount of water required for the curing process and also possibly on the concreting speed of machines using sliding formwork.

Our view is that these cements involve certain inconveniences with regard to canal construction, but may be justified when, after a thorough study, the suitability of their use on ground containing gypsum can be verified.

Another sulphate-resistant cement, which is based on a completely different idea, is the so-called "aluminous cement". Its initial components are limestone and bauxite, which are combined at a temperature of 1500 °C. It possesses the peculiarity of not producing free lime on setting and hence it cannot then be attacked by gypsum and does not form any Ettringite or calcium sulphoaluminate. Instead of free lime, it releases acidic aluminium hydroxide.

Although this cement is completely resistant to sulphate attack, it has other characteristics which, in our opinion, do not make it recommendable for canal construction.

It is a type of cement that does not set well at high temperatures above 25 °C. In all irrigation canals, it is normal for even higher temperatures to be reached, in which case, it should not be employed.

In addition, its acidic nature means that the protection of reinforcement steel is lost, the main defence of which is the basic character of the free lime. It must not therefore be used in reinforced concrete structures. There have been some significant and extensive failures in Spain through the use of this cement. It has been employed a great deal in prefabricated structures because it hardens much quicker than ordinary Portland cement, it can attain significant strength within only 24 hours.

Since this property is not important in canals, we can safely state that this type of cement is not recommendable for their construction.

We shall sum up by stating that the use of normal Portland cement is valid for canal construction, but with a limit placed on the amount of tricalcium aluminate. If gypsum is present in the underlying ground then Portland cement containing bicalcium silicate as the basic component should be used, with very little tricalcium silicate and tricalcium aluminate.

Neither siderurgical nor aluminous cements are recommended, but pozzolanic may be considered where significant problems of gypsum in the ground exist, but knowing well that it will be necessary to maintain the concrete wet during a much longer period of time.

The lining thickness tolerance, in relation to the design thickness is 10% less at all points, provided the average does not exceed 20% of the theoretical thickness.

There are important codes available for concrete (77), which should be strictly followed.

Chapter 6

Prefabricated slab and brick linings: Rubble-work linings

6.1 Prefabricated slabs

The concrete that has been referred to in previous chapters was made onsite, which means that it was mixed, poured, extended and compacted next to or over the actual canal.

It is well known that concrete can be used to produce high quality prefabricated pieces, which can also be employed to line canals.

The method used is that of employing thin slabs, which is intended to help reduce both transport and installation costs. The actual dimensions are very variable, from 30 cm × 50 cm to 2 m × 1 m.

The material's strength depends on its thickness and this has to permit the slabs to be transported, raised and installed over the trimmed canal excavation. Of course, if there are any protruding points in the excavation (though everything must be done to avoid this), the slabs should be strong enough to withstand the water pressure without breaking.

It must also be taken into account that the maximum aggregate size in the concrete, in the case of lining and piping, should not exceed one-third of the total lining thickness. This will greatly limit the maximum gravel size and could lead to high cost increases that might negate the original savings due to the lower volume of concrete required.

Perhaps the greatest advantage associated with prefabrication is the excellent product quality that is obtained, which is much better than concrete that is mixed onsite, and it is also less difficult to control and involves fewer weather-associated problems for the people working with it, all of which lead to better overall quality results. The concrete production plant is able to provide much better concrete control with laboratory and inspection techniques than when it is mixed and poured onsite, since these will be separated by large distances. The concrete must possess very high strength and so great care must be taken, with respect to not only the aggregate size, but also the water–cement ratio, which should be as low as that permitted by the available means of compaction.

In order to achieve lower water–cement ratios during slab production, powerful vibratory benches are employed, some of which include suction systems to remove any excess water. Formwork that incorporates close-woven fabric on the upper slab surfaces is used and this enables the elimination of excess water, but without allowing the cement to pass and results in water–cement ratios that are impossible to achieve under normal production conditions. The strength of the concrete produced in this way is extremely high.

This, then, is the main reason why, under normal circumstances, thin slabs can be achieved without any requirement for reinforcement or prestressing.

Another significant advantage for certain climate regions is that it now becomes possible to take most advantage of the times of the year when, due to adverse weather conditions, production would be otherwise very low. This would be the case, for example, in regions where there are very well-defined wet and dry seasons. During the rainy season, it would be possible to prefabricate a large number of slabs and store them under some form of suitable cover that would protect the personnel involved from the adverse weather conditions. Then, on arrival of the dry season, large surface areas of lining could be installed employing simple machinery basically consisting of lorries and small cranes for raising the slabs. This would obviously involve taking precautions to prepare accesses and to carry out the excavation profiling or trimming.

In order to obtain maximum benefit from this method, the associated canal must be large enough to justify and amortize the installation of a prefabricated slab production plant, which would normally be sited in a location with all-round easy access and electric power.

To facilitate this amortization in situations where not just a canal is to be constructed, but a complete irrigation network for a region, including the execution of smaller canals, it is recommended that the entire network be prefabricated. In this case, it is recommended that the contents of Chapter 18 concerning such small canals and irrigation channels be taken into consideration.

One basic characteristic of prefabricated slab linings is that they have a joint length per square metre that is much greater than in the case of onsite concrete. This is still true even with very large slabs, and even more so in the case of smaller ones (Figure 6.1).

The result is that the probability of filtration is much greater in a slab-lined canal. This means that any decision should only be made after very careful analysis, with special attention being paid to whether the canal is basically being constructed with small infiltration. If, in spite of everything, it is decided that a prefabricated slab canal is the most recommended, extreme caution must be taken with the joints between slabs in order to obtain maximum impermeability guarantees, otherwise it might turn out to be difficult to achieve desired losses of less than 25–$50 l/m^2$, unless, of course, the underlying ground itself possesses sufficient level of impermeability.

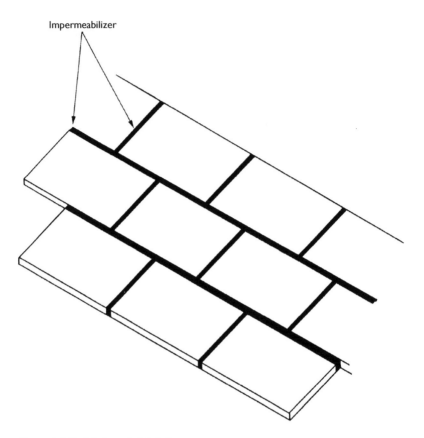

Impermeabilizer

Figure 6.1 Prefabricated slab lining.

A very different case is that of a canal that is lined to prevent ground erosion, facilitate its cleaning, and improve the rugosity factor and hence the canal capacity, etc. The profusion of joints in this case should not be a problem.

The quality achieved in the excavation trimming or profiling has tremendous significance in the use of prefabricated slabs in canal linings. Not only must there not be any irregularities in the ground that could possibly break the slab, but the excavated ground plane must match the desired surface. However, the most important point is that there must not be any excessive excavation at any point. If this were to occur, then the hollow should be filled with low-cement concrete since it is completely unacceptable to use soil for this purpose, which would not be suitably compacted, as can be seen in Figure 6.2.

Figure 6.3 shows a clear example of how an excavation should not be carried out where prefabricated slabs are to be installed.

Figure 6.2 Unacceptable slab installation.

Figure 6.3 Unacceptable excavation for prefabricated slabs.

The canal slopes should be gentle in order to guarantee slab stability due to ground friction. Otherwise, any breakage of any one of them would lead to the undesired movement of an entire zone.

Under normal circumstances, the slabs are installed so that each one is supported on two others in the lower row. This will produce an overall assembly that is beneficial to stability.

A horizontal rim edge is often fitted to the top of the lining using the same type of slabs, which, in addition to providing an aesthetic finish, assists in preventing rain water entering via the linings extrados. This can be seen in Canal V in the Orellana Perimeter, shown in Figure 6.4.

The large number of joints per unit of slab surface area means that the type of joint employed should not be excessively costly. Several systems have been used, but perhaps the most logical choice (Figure 6.5) is to use cement mortar for the joint interior, directly over the ground, together with one of the impermeabilization materials described in Chapter 4 over this. The mortar will provide a certain mechanical strength, which would support the overlying slab weight if this were to slip, with the other material providing the necessary impermeabilization.

This type of lining does not require any contraction joints because there is no setting shrinkage and if there are any periods of low temperature, the involved contraction would be absorbed by the very large number of construction joints between the slabs.

Figure 6.4 Canal V – Orellana Perimeter.

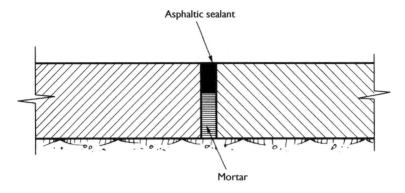

Figure 6.5 Joints between slabs.

6.2 Brick linings

Ceramic bricks produced from fired clay are employed in certain parts of the world where suitable clay deposits are abundant, to construct canal linings that are very similar in characteristics to prefabricated concrete slab linings.

They are laid using cement mortar, often on top of a layer of mortar extended over the ground, which will improve location in the canal plane and the impermeability. Mortar is also used between bricks, both horizontally and vertically, which will also endow a certain level of impermeability.

Very occasionally, a second layer of bricks is laid over the first to improve the lining quality. This is how the Haveli Canal in India was constructed. A layer of approximately 1 cm of mortar, mixed with one part cement and three parts sand, was placed between the two layers of bricks (Figure 6.6).

Also 6-mm iron bar reinforcement was employed, separated by some 30 cm, to create a mesh. This mesh was eliminated in subsequent projects because problems arose in situations of uplift pressure that cut off the water exit from behind the lining, which otherwise would act as drainage.

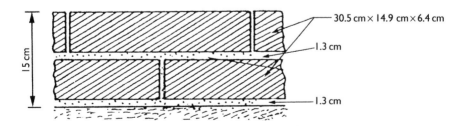

Figure 6.6 Lining with two layers of bricks.

Figure 6.7 A small brickwork canal in Argentina.

No transversal or longitudinal joints were employed because the numerous cracks that appeared in the mortar between the bricks due to shrinkage made them unnecessary.

It should be evident that because of the smaller size of the brick in relation to the prefabricated slabs, and the lack of joint impermeabilization treatment, the level of infiltration was higher. The actual values achieved over permeable ground were considerably worse than the desired ones of $25-50 l/m^2$ in 24 hours. However, this type of lining can be useful under certain circumstances.

It should also be quite clear that this system is extremely labour intensive, which means it can only be recommended in regions where this is cheap or for very small canals. Figure 6.7 shows a small canal in the Mendoza region of Argentina in which bricks were used to construct the side walls.

6.3   Rubble-work linings

Rubble work is a construction that consists of large pieces of stone or rubble that are positioned by hand, with any cavities between them being filled with cement mortar.

Some of its main characteristics are its high level of surface irregularity, which means a poor rugosity factor and the much thicker linings it produces, together with insufficient impermeability. All this indicates a type of lining that can only be used under very special conditions, such as those constructed over impermeable rocky terrain, or for evacuation or drainage canals.

One situation where this type has been successfully employed is for the first sections of a canal with a direct connection to a river, in the area where there is a ground water layer that is greatly fed by the river. The great

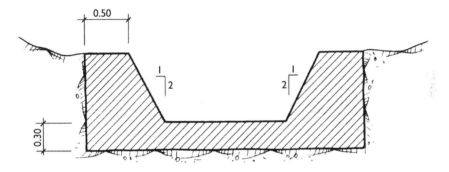

Figure 6.8 Possible cross section with high ground water layer.

Figure 6.9 A small canal in Argentina.

danger for such canals is the permanent uplift pressure that could lead to them floating when empty. The great mass of the rubble layer would ballast the canal in this situation and prevent this from happening (Figure 6.8).

In general, even in zones where rock is abundant, this type of lining involves high cost because of the amount of labour required and the resulting large volume of construction. It is sometimes used for small canals, where the use of stone can resolve the search for construction materials, as shown by the small canal in Mendoza, Argentina (Figure 6.9).

Chapter 7

Asphalt linings for canals

7.1 General characteristics

The basic characteristic of mass concrete canal linings is perhaps their great rigidity, which makes them highly vulnerable to cracking caused by differential settling of the underlying ground or due to heat stress.

On the other hand, there are some types of canal linings that can be considered as being somewhat flexible because they posses a certain capacity to deform without cracking. Prefabricated and brick linings are sometimes known as "semi-flexible" because their numerous construction joints adapt to the ground better than linings that are constructed onsite.

Typical flexible linings are the so-called bituminous or asphalt ones and in their most widely used form consist of a thin layer of a mixture of bitumen, gravel, sand and powder, also called filler, all of which are duly compacted. This mixture is also known as "asphalt agglomerate" and even "asphalt concrete" because it has a composition that is similar to concrete, however the binder in this case is bitumen in place of cement. It is also different in the fact it does not include any water and the much smaller aggregate.

In this text, bitumen is a product obtained from asphalt through a process of fractional distillation, which has certain penetration powers (see Section 4.4). Bitumen can also be obtained from petroleum.

Another basic characteristic differentiating between flexible asphalt and rigid concrete linings is that the former possesses great capacity to withstand deformation and thus adapts well to variations of temperature, lengthening or shortening as necessary with very small compression or traction stress, which means that these types of canals do not require joints.

Nevertheless, they do have significant disadvantages. If there is excessive heat from sunlight when the canal is empty, there is a danger that the lining could soften sufficiently to tend to flow to the bottom and the lower parts of the slopes.

These linings are softer than those of cement concrete and do not withstand the trampling of livestock when they drink from the canal.

Perhaps the biggest disadvantage comes from its lack of resistance with respect to the growth of vegetation which are able to easily perforate the linings at the beginning and then end up destroying them.

Although the linings constructed with asphalt concrete are the typical type of asphalt linings, at times very thin, membrane-type, impermeable linings are employed, which are also made from asphalt. The material's extreme thinness must also be summed to the previously described disadvantages, which obviously makes them more vulnerable, and it is necessary to further protect the lining with a layer of fine grain material or reinforce the membrane by some suitable method. It should not be forgotten that water in movement possesses an indisputable erosive capability.

For this reason, these types of linings have, in our opinion, greatest application in the impermeabilization of tanks and ponds and are a very economic solution in such cases.

7.2 The fight against vegetation

Since vegetation is the first and foremost enemy of bituminous linings and dangerous to plastic membranes, special attention should be paid to this problem during the project and also during the execution of the works.

In general, the growth of vegetation is checked by the addition of some type of herbicide to the ground immediately before the lining is put in place.

Of course this treatment has to be carried out without reduction in the essential condition that the underlying ground must be well compacted and free from roots and any vegetable cover.

In all canals, especially in those for irrigation or supplying drinking to townships, extreme caution must be taken to ensure that no poisonous substances are employed, which could first contaminate the circulating water and then agricultural products or drinking water, with very serious consequences for the population's health.

This quite obviously means that all products containing arsenic compounds are completely out of the question, even though certain publications describe them as being useful herbicides. A mixture of borax and sodium chlorate has also been described as very efficient (25). Several authors recommend a proportion of 10 to 4 by weight. The application dosage is approximately $250 \, gm/m^2$. Great care should be taken when handling sodium chlorate because it is highly inflammable, and a study should also be performed in the case of irrigation canals to ascertain whether the boron involved will be harmful to the crops that are to be irrigated.

In recent years, special products have been announced (25) which contain special substances dissolved in heavy oil, for example pentachlorophenol, in a proportion of 1.5%. The stated dose rate is $3–4 \, l/m^2$, however, never in a single application, but in 3 or 4 doses separated in time. This will enable the product to enter the ground better and not to slide down the slopes.

The fact of the matter is that it is a difficult problem to fully resolve because a herbicide is required kill the vegetation, but should not be harmful to humans, livestock or the actual crops being irrigated.

If the canal is for hydro-electrical use, there is still the worry of avoiding contamination of the river waters further downstream.

7.3 Asphalt concrete linings

These linings have the best quality from among those examined in this chapter.

The asphalt concrete (or hot mix) is a mixture of gravel, sand, inert powder, which is generally known as "filler", together with bitumen. This is invariably laid hot and then compacted until it only has a small void index resulting in an excellent level of impermeability. The tolerance band corresponds to what is known as a closed granulometry, which indicates excellent compactness.

The lining thickness normally varies between 2.5 and 5 cm with others being possible but infrequent. They can be constructed in a single layer or two superimposed layers in which case whatever individual layer sizes are required may be selected. Each layer thickness should not exceed a maximum thickness of 4 cm in order to achieve good compaction, nor less than 2.5 cm for fitting and economy reasons.

These canals are constructed without any contraction or expansion joints, they only have construction joints. These offer greater impermeability guarantees when double layer linings are used by simply arranging for the upper layer to be positioned over the construction joint of the lower layer.

Whatever the case, it is necessary for contiguous layers to be solidly welded together, especially in a single layer lining. On occasions, this is achieved by heating the joint zone, or by means of bituminous paint, which is applied on top of the construction joint.

The bituminous binder should have a penetration that is suitable for the temperatures of the place where the canal is to be constructed and should be justified by means of experimentation and trials.

However, bitumen with a penetration of 60/70 can be used for the slopes, although this would have to be reduced in hot climates to 40, for example. For the bottom, where compaction is easier, bitumen with a penetration of 90/100 may be employed. We would like to remind you that bitumen test piece penetration is measured with a specifically weighted needle at a predetermined temperature. The softer the bitumen, the greater the level of penetration.

The correct quantity of bitumen is essential if good impermeability is to be obtained, together with sufficient aggregate bonding. However, an excess is counterproductive since this would facilitate the sliding of the mixture down the slopes at high temperatures. The correct quantity lies between 8 and 10%.

Calcareous aggregates are preferable because they usually possess greater adherence to the bitumen. The gravel size should be no greater than one-third the required lining thickness to facilitate the compaction operations.

It is very important to obtain a closed aggregate granulometry in order to achieve improved stability and impermeability. The American Asphalt Institute recommends the following granulometric formulas for bituminous agglomerate, which may be used as guideline values for the definition of the mixtures to be employed.

Mixture number	I	II
Minimum thickness	3.8 cm	2.5 cm
Sieve	Percentage by weight	
3/4 in.	100	–
1/2 in.	95/100	–
3/8 in.	–	100
No. 4	60/80	90/97
No. 10	40/55	65/80
No. 40	25/35	35/47
No. 80	15/27	22/30
No. 200	8/15	10/16
% bitumen by weight	7/9	8/10

One very important characteristic of agglomerates (asphalt concrete) used for canals is the level of impermeability. To this end, low percentages of small sand are sometimes added to the mixture, with sizes between 0.1 and 1 mm.

Workability will be improved if the aggregates are rounded.

The high density of the mixture, assuming correct dosage as described earlier, is obtained onsite with good compaction at a sufficiently high temperature.

Many works that require test pieces compacted in the laboratory, with a pressure of $20\,kg/cm^2$ and $140\,°C$ mixture application temperature, have a void percentage that does not exceed 2.5%, and in general with permeability Darcy factors of $10^{-7}\,cm/s$, equivalent to $10^{-9}\,m/s$.

One important point is that the ground should not be damp when it is applied.

Asphalt lining repair work is much simpler because the materials involved join well together when heat is applied.

There are many factors that have an effect on the quality and characteristics of an agglomerate lining. For example, a greater content of asphalt, or in other words, bitumen, will increase the impermeability and reduce slope stability. It also reduces the erosion resistance, but increases workability. A greater content of aggregates (gravel) will increase slope stability and erosion resistance, but will decrease workability.

However, from among all the considered variables, works execution, compaction pressure and mixture temperature take pride of place, with all three having great importance in obtaining a good level of compactness and hence good impermeability.

Figure 7.1 has been included for examination. It consists of two diagrams. In the right-hand side, there are two cartesian axes that represent the compaction pressure and the void percentage. There are two curves: one corresponding to the compaction temperature of 110°C and the other to 140°C. By establishing the compaction pressure on the abscissa axis and moving vertically until it meets the curve, one point is established. The horizontal to the vertical axis defines the porosity percentage.

There are other cartesian axes in the left-hand side, in which the ordinate axis represents porosity and the abscissa the Darcy permeability k of the bituminous lining, in cm/s. There are also two curves corresponding to the same temperatures of 110°C and 140°C. By commencing with the value obtained for the porosity and moving horizontally to the temperature curve, the permeability is obtained.

In Section 2.7, we saw that to obtain the desired impermeability for a well-designed canal, a 10 cm thick clay lining (or other lining of equivalent imperviousness) was required, with permeability factor of 10^{-6} cm/s (10^{-8} m/s), which is obtained with normal or equivalent clays, and since the thickness of asphalt mix usually is half this amount, then half the permeability factor is required also. To include a certain safety factor, a value of 10^{-9} m/s (10^{-7} cm/s) is usually employed.

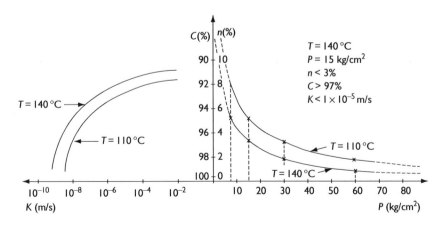

Figure 7.1 Permeability of asphalt concrete.

AN EXAMPLE

With a compaction pressure of $30\,kg/cm^2$ and a compaction temperature of $110\,°C$, an approximate void percentage of 3.5% is produced, together with a permeability of $10^{-8}\,m/s$, which is considered unacceptable.

On the other hand, if the compaction operations are carried out at $140\,°C$, with the same compaction pressure, the desired permeability level of $10^{-7}\,cm/s$ is achieved and the void percentage is 2%.

7.4   The shape of the canal's cross section

In this type of canal, the cross section has slopes that are defined not only by their condition of being stable, but also because they provide greater safety with respect to lining slippage. This can be a serious danger in sheets, even in the case of asphalt agglomerate, when temperatures rise and the canal is empty.

The choice of slopes also plays a significant role, together with the constructional requirements, for example, to prevent bituminous mixture creep.

It can be said that, in general, these canals have slopes lesser than those of cement concrete-lined canals, for example, inclinations of 1.75/1 or 2/1, or even less.

The joint angle between the slopes and the bottom should not be acute because the same problem of achieving sufficient impermeability as for cement concrete linings would occur. Instead, it should be rounded so that it does not lead to any problems during construction, even when the extending and the compaction is carried out in longitudinal direction. However, in general, except where the canal is large enough, the compaction is performed in transverse sections, with a roller that is raised and lowered along the slopes by means of a winch installed at the top of the transverse section. A circle having a radius that is not less than half the bottom width, with a minimum of 1 m, is usually adopted as the joint curve in order to facilitate the passage of the compaction roller.

The rugosity of these linings is comparable to that obtained with cement concrete and well-designed and constructed joints.

The permitted tolerances for these canals are the same as those for cement concrete linings, which are 5 cm in plan errors for straight alignments, 10 cm in curved alignments and 2 cm in heights, all measured with a 4 m long ruler. It is normal to require that the minimum thickness does not have any errors exceeding 10%.

It is essential for the excavation to be well trimmed since any small errors can lead to very significant percentage errors in the volume of agglomerate used. Any areas of excessive excavation should never be filled in with compacted filling, but instead with low-dose gravel cement with, for example, 80–100 kg of cement per cubic metre. They can also be filled in with asphalt agglomerate, with a low dose of 3–4% of bitumen. The latter method is recommended when it is necessary to place a layer of filter material underneath the actual lining so that it can act as drainage. In this case, the low-dose layer serves both purposes. The definition of filter layer thickness is a very interesting problem that is common to many types of linings and is covered in greater detail in Chapter 9, which specifically deals with these matters.

7.5 Asphalt membrane linings

The very first membranes that were used for linings consisted of a hot-melt bitumen layer, which was extended over levelled ground that had been well compacted and cleared of all vegetation.

Six litres per square metre were applied and since the density of bitumen is close to 1, the membrane thickness was some 5–7 mm.

This membrane provides sufficient impermeability when applied to correctly prepared ground. Nevertheless, it is extremely weak with respect to sunlight, water erosion and treading by livestock, etc. This therefore requires a total covering of between 30 and 60 cm, divided into two layers, where the lower one is soil or sand and the overlying one is gravel.

This protective layer has to be made with painstaking care so that it is well extended in spite of its thinness and, what is more difficult, without damaging the bituminous membrane in any way.

From the hydraulic point of view, this type of canals behave in the same way as unlined canals, which means that all the erosion, sedimentation, rugosity factor and maximum speed studies, etc. have to be carried out and which we will look at in Chapter 11. All this makes it even more difficult to produce good results with such a poor quality impermeable layer and for this reason it is only recommended for the construction of ponds, and not for canals. If a membrane lining is to be used, whether protected or not with granular material, the prefabricated asphalt products listed below should be employed or the plastic materials that are described in the next chapter.

In order to obtain better quality membranes, prefabricated asphalt membranes have been used on occasions, consisting of an artificial fibre core soaked in bitumen that has usually been oxidized, whereby air or oxygen was bubbled through while it was molten. This endows the bitumen with improved properties with respect to simultaneous resistance to both high and low temperatures.

These membranes may incorporate several superimposed layers that are combined together to produce the desired thickness. They are normally manufactured in lengths from 5 to 10 m, with widths of around 1 m and thicknesses between 3 and 10 mm.

The enhanced strength of this type of membrane enables them to be employed with or without protective surface layers.

Construction joints are produced by overlapping each strip with the next by some 6–10 cm and then applying heat to weld them together. They can also be laid without any overlap if a hot asphalt seal is applied to the joint.

Chapter 8

Plastic membrane linings

The idea of using thin bituminous membranes has also led to the use of plastic sheeting, which is completely impermeable and many types possess excellent tensile strength.

All plastic membranes, whether bituminous or plastic material, are even more flexible than the asphalt agglomerate lining described in the previous chapter.

One of the basic advantages of plastic sheeting over the previously described bituminous membranes is that some of them possess much greater resistance to erosion and puncture, and do not undergo any softening with heat, so that there is no danger of any loosening similar to that of bitumen softening.

In the same way as bituminous linings, plastic membranes can also be seriously affected by the natural growth of vegetation, which means treatment of the underlying ground is required prior to laying the membrane. This treatment should be carried out with good quality herbicides, together with thorough ground clearing, eliminating all roots and other organic material. However, there are many plastic membranes that possess great resistance to puncture so that if in spite of all precautions a certain amount of vegetation is able to grow, the material will be able to withstand it within certain limits.

There are many types of plastic materials available, but the most widely used are polyvinyl chloride (PVC), high density polyethylene (HDPE) and butyl. PVC sheets are usually much thinner than butyl ones.

Nevertheless, plastic membranes are available on the market, which have various thicknesses, which, in some materials, can be as thick as 3 mm.

It is evident that their strength depends not only on the type of material, but also on its thickness. This leads to two different types of plastic membrane linings: those consisting of thin sheets, together with a protective layer of gravel and sand; and thicker sheets without any protective layers and in direct contact with water. Each of these possible solutions shall be described separately.

8.1   PVC membrane linings with protective gravel layers

The usually employed PVC is manufactured with a plasticizing agent, having taken into consideration all the possible problems associated with aging. It must also be taken into account that the loss of the good characteristics of plastic materials over time is perhaps their most significant disadvantage. When these materials are employed, not as sheets, but in the manufacture of resistant elements, such as piping, the extrapolated strength expected after fifty years, which is significantly lower, is used (66). When employed in sheet form, they also lose some of their good strength and flexibility properties, and should be used with suitable safety factors.

This, without any doubt, is one of the reasons why North American engineers, in a country where plastic sheeting has undergone a great deal of study, have increased the minimum recommended PVC thickness from 10 thousands of an inch (0.24 mm) to 20 thousands of an inch (0.48 mm).

The number of available standards, above all, North American ones, referring to the verification of the characteristics of this type of material is very large (1), (7), (9), although there are very other interesting ones, which are usually based on them. In general, the required conditions are as follows:

The tensile strength that the membrane is required to have in both directions is 8 N/mm (8 kg/cm) for the 20 thousands of an inch thickness. The minimum elongation has to be 300% and the modulus of elongation, again in each direction, must be a minimum of 3.2 N/mm.

The tensile strength at the joint between two consecutive sheets must be a minimum of 80% of that of the sheet.

The plasticizing agent stability, as measured in accordance with the ASTM D1239 standard (1), immersed in distilled water at 50°C for 24 hours, will have a maximum 0.35% water loss. It will also have a maximum 0.9% loss of volatile substances, as measured by the ASTM D1203 standard, for sheets with a thickness of 20 thousands of an inch.

Buried sheets are able to achieve excellent levels of impermeability, providing the materials, design and construction are good.

The described sheet, however, will only have poor resistance to external agents and will, therefore, be covered by a layer of granular materials. The purpose of this layer is to hold the sheet in place, protect it from water velocity, wave and wind action and also perforation by roots and livestock hooves.

From a hydraulic operational point of view, the canal will behave like an unlined canal because of the protective layer, with a Manning coefficient of between 0.0225 and 0.025, or more accurately:

$$n = 0.028c \cdot (d50)^{1/6}$$

where $d50$ indicates the sieve aperture in metres that retains 50% of the material by weight (27).

This requires that the canal has a water velocity that does not erode the protective layer as will be examined in Chapter 11, which can influence not only the slope but also the transverse section. In any case, the slopes should be sufficiently gentle not only for the sake of stability, but also with respect to sheet and protective layer stability, both during construction and throughout the canal's lifetime. This will require a minimum of a 2/1 slope. Sharp angles must be avoided between the canal's bottom and its slopes.

The protective layer should have a minimum thickness of 40 cm, which means the excavation should be increased by this same amount. The material is normally laid by means of a clamshell excavator and the profile trimming should be carried out with great care, also usually employing a clamshell excavator.

At times, when the excavated area shows irregularities, all hollows should be filled in with some 10 cm of sand or fine grain soil in order to smoothen the membrane substrate.

With respect to the sheet's protective upper layer, two of these can be laid. The first being made up of small-diameter material, usually sand, with the second consisting of larger-diameter material. The purpose of the first finer layer is to protect the membrane, while the second is intended to withstand the water's erosive force. The permeability of the upper layer should be greater than that of the lower one to prevent uplift pressure. The total thickness may be some 40–50 cm. However, there have been certain cases where the finer lower layer caused the upper layer to slide downwards.

Figure 8.1 shows a cross section of a canal with a buried membrane lining.

Polyvinyl chloride membrane sheeting is usually purchased in rolls of 100 metres. The ends of consecutive rolls are joined by means of construction joints, which has to be carried out with care. Although PVC sheets tend to weld themselves to each other with the passage of time, they still have to be initially joined by applying a solvent supplied by the manufacturer. This

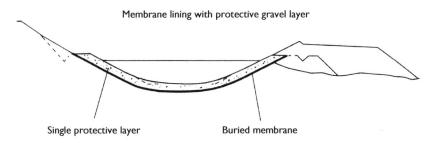

Membrane lining with protective gravel layer

Single protective layer   Buried membrane

Figure 8.1 Cross section of a canal with buried membrane.

should be applied to an approximately 5 cm overlap zone of one sheet over the other, which should be of total width of at least 30 cm.

If polyethylene is used instead of PVC, the welding operation becomes more difficult. It is still carried out in the same fashion, but in this case employing adhesive polyethylene tape.

The greatest difficulty occurs during windy weather. The laying operation should be carried out in the same direction as the wind, if at all this is possible.

The sheeting should be weighted down with the same material that is to be employed as the protective layer as the laying operation advances. The ballasting has greatest importance at the upper ends of the canal banks since the sheet always has a clear tendency to slide towards the bottom.

The protective covering material should be applied as soon as there is a suitable section of canal with correctly laid sheeting. The material is not compacted, because the sheeting has to be protected from puncturing, although trimming is often required in order to obtain the necessary thickness.

The specifications that are usually imposed by the protective material in a single layer are as follows:

Size through	Percentage passing
75 mm	90–100
38 mm (1.5 in.)	80–100
19 mm (0.75 in.)	65–100
9.5 mm (0.375 in.)	40–95
4.75 mm (sieve No. 4)	30–85
2.36 mm (sieve No. 8)	20–70
1.18 mm (sieve No. 16)	10–55
Sieve No. 50	0–20
Sieve No. 100	0–10
Sieve No. 200	0–5

This is the desired granulometry, according to North American publications, for aggregates with a maximum size of between 1 and 3 in. (between 2.5 and 7.5 cm), but it must not be forgotten that this has to be achieved as cheaply as possible, which means that crushing, with subsequent classification and mixing, is not recommended unless it is absolutely essential.

Anyway, it has to be pointed out that in the USA they do not have a very clear idea as to the size to use, with these being different depending on whether one or two layers are employed (27).

In a significant paper by the Bureau of Reclamation (27), several different sizes are examined for the protective covering layers used in several North American canals employing membranes. The results widely differ, and although several granulometry sizes were finally proposed for the case with two protective layers, we have preferred to provide the example of the granulometry used in the Amarillo Canal in the USA:

Size	Percentage
Fine lower layer	
2 mm	100%
1 mm	100–98%
0.3 mm	95–90%
0.1 mm	55–25%
Sieve No. 200	12–40%
Course upper layer	
5″ (12 cm)	95–80%
3″ (7.2 cm)	85–65%
1.5″ (3.6 cm)	70–45%
0.75″ (1.8 cm)	55–35%
2 mm	40–20%
0.3 mm	30–12%
Sieve No. 200	5–0%

8.2 Unprotected membrane linings

There are several types of materials employed as membranes without any overlying soil protection, in other words, those directly exposed to the water in the canal or to sunlight when it is empty (known as exposed membranes), all with the very necessary property of being sufficiently resistant to them. In general, all that is described in the following sections can be applied to almost all plastic membranes with greater thickness and strength than the thin PVC sheeting of the previous section.

The philosophy behind these linings is intended to eliminate the need for the expense of protective gravel or sand layers thanks to the greater strength of the plastic membrane.

The most widely employed materials are butyl and other synthetic rubbers (51), together with HDPL.

Butyl is obtained through the combining of two petroleum derivatives, isobutylene and isopropylene, at low temperatures. This results in a highly flexible synthetic material, which is resistant to heat, cold, light and ozone. It always maintains a high level of tensile strength, and is puncture resistant, highly impermeable and very flexible.

High density polyethylene is also flexible and includes carbon black in its mass to provide protection against sunlight. There is a certain amount of experience of its use in Spain, mainly in pools and small reservoirs (19).

There are both advantages and disadvantages associated with the use of exposed membranes. They include the following advantages:

- They can be employed with steeper slopes (1.7/1), provided the membranes are sufficiently anchored at the berms or upper sections of the canal, since sheet loosening is an ever-present problem, but there no longer is the need to comply with the much more demanding stability conditions of the covering material.
- Although of lesser importance, the extra excavation work is not required with respect to the extra depth necessary for the protective layers.
- Any faults that are present are easily detected and located.

The disadvantages are:

- Damage caused by livestock hooves
- Erosion caused by water
- Aging due to sunlight
- Sheet movement due to the effect of water and/or wind when the canal is empty
- Vandalism.

In our opinion, the last one is the most serious. In tropical countries, we have seen cabins with straw roofs that were made waterproof by installing a piece of plastic membrane that had been cut from a canal lining with a machete normally used for cutting sugar cane. Just this one fact makes us feel that this type of lining may be suitable for large private properties where there is permanent efficient vigilance, and it is not, however, recommended in areas of public irrigation.

There are usually high water velocities in hydroelectric canals that could raise the membrane and this makes it a very significant point for prior study.

One important danger in exposed membrane linings is the possibility of sliding towards the bottom. The only method of fixing them in place is to anchor it to the upper sections of the slopes. This requires the construction of a small trench, some 25 cm deep and wide, along the full length of the canal. The upper section of the membrane is placed in this trench, which is then filled in with soil and compacted (Figure 8.2).

30 cm × 30 cm

Figure 8.2 Sheet anchorage.

In an auxiliary section of the Lodosa Canal, which had been lined with 2 mm thick HDPL, the sheeting was anchored to the bottom and to concrete beams on the upper canal berms. This solution, from the sheet anchoring viewpoint, would seem to be quite satisfactory (18) (Figure 8.3).

Excavation trimming should be carried out very carefully to ensure that no sharp stones are left. If the ground is rocky or otherwise irregular, it is recommended that any hollows be filled in using some 10 cm of selected material.

However, since the puncture resistance for butyl sheets that are 0.79 mm in thickness is $2.81 \, kg/cm^2$, the behaviour of the sheets in general is quite satisfactory.

Butyl's resistance to ozone varies a great deal depending on the manufacturer. There were cases in tests carried out at 38°C, with 20% stretching, where the test pieces were able to withstand for one hundred days without any alteration, and others where they failed after only seven days, which is the minimum period required by the Bureau of Reclamation in their specifications sheets. On the other hand, the artificial rubber EPDM is completely resistant to attack by ozone, which gives it a great advantage.

With respect to the test where the material is left to the ravages of outside weather, both the butyl and the artificial rubber remain in good condition after periods ranging from four to twelve years depending on the case. Once again it must be pointed out that the aging problem is one of the most serious involved with these types of materials. To establish their real costs, it would be necessary to take into account the required period between two successive replacements.

The commercially available thicknesses for butyl membranes are 0.77 mm, 1.5 mm, 2.25 mm and 3 mm. The 0.77 mm thickness is sufficient for canals and irrigation channels, although the current trend is to employ ever-increasing thicknesses.

Cross section of the Lodosa Canal

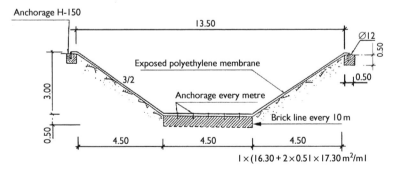

Figure 8.3 Sheet anchorage in Lodosa Canal.

Butyl sheeting is sold in rolls of up to 50 m with maximum widths of 12 m, and even larger dimensions are available on special order. They can, therefore, be laid either longitudinally or transversely as required. Whatever the case, some thought must be given to the joints. The best method is again, by overlapping the two edges and employing a solvent supplied by the manufacturer, with special care taken to ensure the upper sheet is placed so that it cannot be lifted by the current of water (Figure 8.4).

A 10 cm strip of another special material is sometimes interleaved between the two ends that are to be glued together, and which possesses better adherence to the two plastic sheets.

The results that were jointly obtained by the Utah State University and the US Department of Agriculture are very significant here. According to their findings, the initial resistance to tearing of the butyl was 125 pounds per square inch (approximately $9 \, kg/cm^2$), which was reduced to 78 pounds per square inch (approximately $5.6 \, kg/cm^2$) after nine-years exposure to outside weather conditions. As always, it was noted that the single great problem associated with plastic materials is aging.

Another material that is employed as exposed membranes for canal linings is high density polyethylene. Figure 8.5 shows the Chongón-Subeybaja Canal in Ecuador, with a 2 mm thick polyethylene lining.

The conclusions obtained from certain Spanish reservoirs are also very significant. The basic difference between reservoirs and canals, apart from the fact that there is no water velocity in the former, is that reservoirs have a much greater depth than canals, which obviously requires greater membrane thickness.

Significant reductions in resistance to tearing were observed five years after construction was completed, but these were only found in zones of strong sunlight and thin membrane, where the losses of 30% in tensile strength and resistance to tearing were noted. The membrane thicknesses used in these reservoirs were 1.54 and 1.74 mm.

We should now be able to deduce that before deciding on constructing a canal based on a membrane lining without any upper protection, it is very necessary to carefully examine the probable durability of the lining and its effects on the true costs of the construction works.

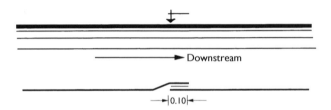

Figure 8.4 Overlapping and flow direction.

Figure 8.5 The Chongón-Subeybaja Canal with polyethylene lining.

However, external membranes can also be very useful for temporary repairs to concrete lined canals, whether prefabricated or concreted onsite, which have fractured. This would be to temporarily prevent water leaks while the irrigation campaign comes to an end so that definitive repairs may then be carried out. In these situations, it is recommended that the membranes be light and manageable, with PVC being the ideal choice.

These have also been employed to carry out longer-lasting repairs to deteriorated canals by anchoring the plastic sheeting to the existing lining or supporting ground. However, much better solution would be to glue the PVC membrane to the fractured concrete lining with an adhesive.

In these cases, it has to be remembered that the concrete lining contraction joints will have significant deformations (Sections 4.2.3 and 4.2.4), and which must be followed by the plastic membrane. Since this membrane is glued to the concrete, only the strip immediately above the joint is able to deform so that its unit deformation could be quite enormous meaning that it will have to be verified that the plastic material is able to withstand it.

At times, filtration in certain deteriorated sections of concrete lined canals can be tremendous (17). Figure 8.6 shows the leaks in the Orellana Canal in 1991.

It can be seen that the leaks in the first 7 km section, which was in very poor conditions, were much greater than in the rest of the canal. This stretch was constructed from prefabricated slabs. The problem was resolved by

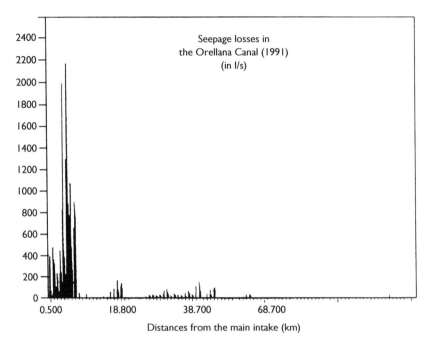

Seepage losses in
the Orellana Canal (1991)
(in l/s)

Distances from the main intake (km)

Figure 8.6 Leaks in the Orellana Canal.

Figure 8.7 Impermeabilization of the Orellana Canal with PVC sheets.

installing 2.25 mm thick sheeting of PVC on top of this lining and anchoring it to the ground (Figure 8.7, taken by Eng. Fermín Jiménez).

It should be taken into consideration that when the sheeting is not glued in place, and merely anchored at the upper banks and transversely section by section, air is able to enter between the sheeting and the lining. The water thrust on the sheeting will drag along the bubbles underneath and transport them, to eventually combine in a large bag of air that could even lift the membrane and drag it free from its anchoring points.

In summing up, all concerned should be aware that plastic sheeting can produce linings that can be very useful under certain circumstances.

Chapter 9

Canal drainage

9.1 The need for drainage

A canal is designed to transport water; however, on many occasions, water is precisely its worst enemy. Yet in this case, the water causing the damage is not that circulating through the canal, but instead, that which filters into the ground.

It can often occur that there are high groundwater layers which, if the canal is emptied, can lead to the lining being uplifted and damaged. Similar effects are often caused by rain water which infiltrates through the ground and on other occasions, the actual canal filtrations through the joints and cracks in the lining.

However, the canal designer has to be able to achieve stability not only of the actual lining itself, but also of the underlying geometric shape, which comes from either the excavation of the natural ground or the backfilled soil. It is a well-known fact that moisture can often produce a lubricating effect in the ground, reducing its cohesion and fostering slope sliding. This sliding can affect the slopes on the side where the canal is located or even the actual side slopes of the canal itself.

However, if the moisture and, even worse still, the underground current lines are harmful to the slopes in excavation, they are even more so for the backfills. It often occurs that the water filtering through a canal lining in a backfilled section has a gradient that is much stronger than a stripping excavation. This can cause higher filtration speeds leading to siphoning and washing away of the backfill, followed by collapsing and serious flooding.

In the ground with more gypsum content, a well-designed drainage system can be an element providing a significant contribution to a canal's stability (Chapter 12).

All this requires the designing engineer to carry out a painstakingly detailed study of the type and number of drains necessary to achieve a high level of probability that the canal will be stable.

9.2 Types of drainage

There is a very wide range of drainage types, which include all those elements that are intended to evacuate any water before it is able to infiltrate into the ground, together with all others designed to eliminate any water that has already infiltrated.

Exterior drainage systems for surface water are obviously the least expensive to construct and also the easiest to maintain. In addition to this, they are also capable of removing much higher volumes of water than interior drainage.

Surface drainage systems consist of ditches, which may be located in the middle of the side where the canal is sited and on a higher level, in which case they are known as "upper ditches". Ditches may also be located on the same bed as the canal or the same backfill bases or even on the sides lower than the canal in order to eliminate infiltrations without causing any damaging effects to the lower zone, in which case they are called "lower ditches".

Interior drainage systems, which are intended to protect the canal lining, are basically piping that allows water to enter along its periphery, then takes it to an exit point where it is evacuated to the exterior. They are sited underneath the lining, housed in trenches filled with filter material which completely surrounds it, along the full length of the canal. This piping may be manufactured using porous concrete, perforated plastic or simply solid clay or concrete, laid with open joints through which the water is able to penetrate. They are always short in length, around 50 cm, but always less than 1 m.

Other useful types of interior drainage are layers of filter material located between the actual lining and the underlying ground. These have exits, to either the exterior or a filter pipe located under the canal bottom.

Among deep drainage elements, which are designed to stabilize not the lining but the canal infrastructure, we can mention some of the most important ones, such as drainage drills in any of their multiple forms and patents, together with filter layers located at the base or middle of the backfills. It is also possible to sometimes design filters at the feet of the slopes.

9.3 Drainage for slope stability

Not only does water capable of damaging a canal come from the actual infiltrations, but also rainwater can lead to instability of canal elements.

Canals constructed in non-rocky ground, especially those intended to provide irrigation, may be located in ground having one of two quite distinct morphologies.

Sometimes the strip where the canal is located is actually a quaternary terrace consisting of sedimentary formations coming from the river plain

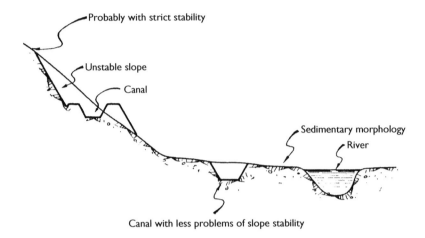

Canal with less problems of slope stability

Figure 9.1 Canals located in different morphologies.

itself, so that the transverse slopes are usually quite gentle because of their origin.

However, above these relatively horizontal terraces lies ground with much steeper slopes due to the fact they were formed mainly by erosion (Figure 9.1).

The former type of ground, because it is created through the sedimentation of successive layers, has an inclination that is much less than that required for stability. On the other hand, it often occurs that the latter types of ground, with their slopes produced by erosion phenomena have strict, or almost strict, stability.

One consequence of this difference is that their behaviours are quite different when referring to the construction of a canal.

In the sedimentary zones of the plain, in general, it is only necessary to take into account the stability of the slopes in the transverse section of the canal, where, due to the relative lack of depth, they do not usually present any really serious problems. By employing a suitable canal layout, there is normally no need to increase the excavation depth very much more than the actual canal depth. There are no stability problems associated with the side slopes.

The only difficulty that normally presents itself on these sections is the high groundwater levels that are sometimes found. There should be no need to state that this should be combated by whatever drainage measures are necessary. However, the drainage problems encountered on plains can often be much more serious than those of erosion sides, because of either very significant water flows or the topographic difficulties associated with its evacuation.

There have been occasions when we have successfully employed canals cross sections formed by thick concrete or rubble work walls, the main purpose of which was to provide sufficient weight to prevent the whole assembly from floating (see Figure 6.8).

On the other hand, canals sited on erosion sides, introduce a slope in the excavation that is frequently unstable. This instability is sometimes immediate, but more often it occurs over time as the land is weathered.

It is well known that a large number of factors are involved in slope stability; however, in general, it is the cohesion that plays the biggest role. Because, in turn, cohesion is a function of the ground moisture levels, if the moisture can be reduced, it follows that stability should increase.

One way of achieving this result would be to dig protection ditches on the upper levels of the side, the purpose of which is to evacuate runoff waters before they are able to soak the area surrounding the unstable slope (Figure 9.2).

The basic advantage provided by these ditches is that they can follow a progression that is topographically completely independent of the canal's progression, so that they can have gradients that are sufficient for rapid evacuation of the collected water.

The distance from the edge of the slope excavation to these ditches has to be equal to or greater than the excavation height. This will produce a partial reduction of the moisture levels in the area where the most probable sliding circle is located.

The convenience of having a ditch on the slope side higher than that of the canal has been recommended on many occasions (27), mainly to eliminate the erosion caused by the trickling of water over the slope. We believe that instead of locating it close to the canal, a much better method would be to site it higher up the side, in an attempt to divert this water before it has the chance to soak the wedge of a possible landslide.

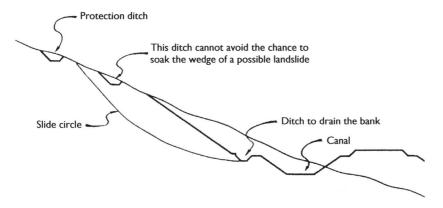

Figure 9.2 Protection ditches.

It is a well-known fact that certain terrains are much more dangerous than others from a stability point of view. According to Terzaghi, heterogeneous ground is among the most dangerous. This type may have more permeable layers that can withhold water, which is then trapped between less permeable layers and can produce very dangerous levels of uplift pressures. In order to improve this situation, on many occasions surface drainage is not enough and this must be aided by internal drainage. This may be carried out in many forms, but one of the most efficient methods is the use of inclined drains normally consisting of plastic drainage piping, although metal is also employed (Figure 9.3).

There are many other procedures for slope stabilization with drainage systems. However, they are often too expensive to be used systematically along the entire length of a canal and are thus only employed at specifically dangerous points that require special treatment.

Surface planting can eliminate erosion and, at the same time, stabilize the slope against certain creep phenomena.

Assuming that the slope excavation is now correct, it is necessary to drain the actual bank itself by digging a ditch parallel to the canal and having a gradient that is equal to or at least very similar to that of the canal.

The purpose of this ditch is to partially free the canal lining from the uplift pressure caused by the infiltrations, which, since they come from the side slope, have not been evacuated by the previous precautions.

This ditch should normally be lined because since it has a gradient that is as small as that of the canal, the water speed is incredibly small because the hydraulic radius of the ditch is much more unfavourable than that of the canal. The lining will improve rugosity and speed.

On the other hand, the lining will prevent filtration from the ditch itself due to the production of swampy areas caused by the plant growth and sedimentation, etc.

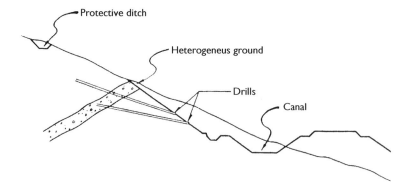

Figure 9.3 Inclined drills to drain heterogeneous soils.

This ditch is usually only located on one canal bank, which is precisely that situated next to the higher part of the hillside, where the service road is not normally sited. The other bank, which is frequently backfilled, does not often have a similar ditch. It should be evident that the inflow of water via this bank is very much smaller and the canal itself carries out the ditch function for the road way, if one exists. Nevertheless, on occasions there is a danger of water entering via the extrados, leading to undermining and uplift pressure. It is frequently considered prudent to apply a layer of tamped clay, which works with the lining to prevent water from entering (Figure 9.4).

Figure 9.4 shows two other terminations that have been employed in Spain for the upper section of the lining. The upper longitudinal beam, which is isolated from the rest of the lining by a joint, is highly useful when the canal is concreted by traditional means. In addition to marking out the alignments and assisting the execution of the various slabs, it can also make it difficult for water to pass since it replaces an area of ground that has been badly treated by the construction works with concrete.

It is quite unacceptable to execute this same beam simultaneously with the lining, but without a joint between the two. The greater weight of the upper edge of the slab would favour the presence of longitudinal cracking due to flexion caused by undermining and ground settlement.

There is another area where surface drainage is necessary, using at least one ditch. These are certain backfill bases where the natural transverse inclination of the side cannot facilitate drainage by itself. We personally went through a very disagreeable experience concerning a canal for which we were responsible. Just as shown in Figure 9.5, the infiltrations within the backfill lead to flooding at its base, which was apparently innocent, but did not have any way out because the side had an inclination in the opposite direction. This flooding was the cause of the softening of the natural ground, which then took on a chocolate-like consistency because the soil was predominantly clay. This phenomenon reduced the ground's natural capacity to withstand the actual weight of the backfill. Successive wedges of this were then left suspended only by the shear force in the vertical planes because the base was resting on pasty ooze. Since the backfill soil was incapable of withstanding such shear force, the successive wedges were eroded and transformed into the same chocolate-like substance once it had fallen. This action went on implacably until some six hours later the backfill was completely destroyed. Fortunately, in a race against time, while this was occurring, it was possible to cut off this inflow of water to this section of the canal by closing off sluice gates further upstream, which prevented significant flooding of the irrigation area.

The existence of a ditch at the foot of the backfill with sufficient inclination to correct the natural ground and achieve a reduction in water content so that it retains enough compression strength is a fundamental requirement.

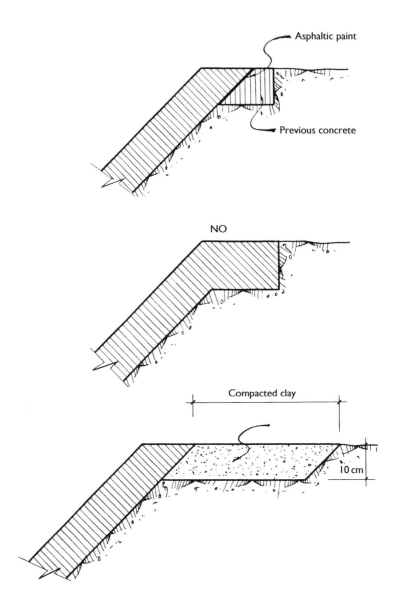

Figure 9.4 Upper part of lining.

This relatively economic measure is very significant with respect to back-fill safety.

There are other drainage methods that can be employed to protect canal backfills, but which are more expensive and not always used. From among them we can mention the drainage layer which is interposed between the

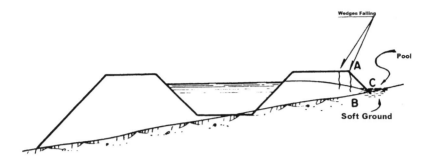

Figure 9.5 Backfill destroyed by drainage problems.

natural ground and the actual backfill, and the filters, which, as in dams, are located at the foot of the backfill slope and are designed to prevent the removal of fines from these locations, which are precisely the lines of infiltration currents.

Of course, these extra measures can be employed in combination with the location of the already described longitudinal drains under the bottom, which will prevent the circulation of filtrations through the body of the backfill, which are fundamentally necessary on occasions for safety reasons.

The danger of infiltration in concrete-lined canals located over high backfills is so great that many engineers consider that this is one of the few situations when the use of interior iron reinforcement bars is clearly recommended. In this way, if the incipient infiltration produces siphoning under the lining, this is able to withstand it just like a slab over the empty space and reducing the magnitude of the leak.

Summing up, the bed slopes where the canal is to be located, whether those of the excavation or those of the backfill, must be examined not only with respect to soil mechanics knowledge, but also taking into account that fact that, in principle, its stability is more precarious than in the case of a highway or railway because of the possible infiltrations.

9.4 Studies of longitudinal drains under the bottom

The basic purpose of filtration piping drains underneath the canal bed is to collect the water that filters through the lining and to transport it to some point where it is able to safely drain away. Its main characteristic is the greater ease of water transport in the direction of the actual drain itself than in the transverse direction.

As stated at Section 9.2, the most economical method of evacuating surface water is precisely through the use of open drains. It is therefore necessary to think of a system in which suitable ditches evacuate the greater part of runoff water, and the drain that is situated under the bottom only has to deal with whatever infiltration exists in spite of everything, which should be minimal if the lining has been both well designed and constructed.

It should not be imagined that these drains under the bottom are capable of reducing the ground water layer in those cases where it too high and too well fed. In this type of situation an adequate drain system must be designed into the project at the same time that other measures are employed to stabilize the transverse canal section and which were discussed in Section 9.3.

The flow rate that normally has to be evacuated by the longitudinal drains under the bottom is precisely the flow rate that filters through the lining. Given that in modern hydraulic techniques, it is considered that the greatest permissible losses in a canal are of the order of $25–50 \, l/m^2$ during 24 hours, this then becomes the design flow rate that should be employed in these circumstances.

However, this entire flow rate, which is the product of this unit loss through the typical canal perimeter, has to be accepted by the unit length of drain underneath, which means that the drain piping perimeter multiplied by its permeability must have acceptable values.

There are several different type of drainage piping available (Section 9.2), which, of course, possess different water admission capacities. Previously used DIN standards required that laboratory analyses carried out on these pipes indicated a permeability or admission capacity of 0.35 l/min and metre length of drain, for an inner diameter of 8 cm and an exterior hydraulic load of some 10 cm.

Since we require that the drainage pipe is capable of evacuating the greatest admissible infiltration in the canal, which we have established as being $50 \, l/m^2$ per day, equivalent to:

$$\frac{50 \, l}{24 \, h \times 60 \, min} = 0.035 \, l/min \text{ and } m^2$$

the result is that a drainage pipe of inner diameter 8 cm is capable of absorbing through infiltration the leaks from a canal, the perimeter of which does not exceed 10 m in its transverse section or 20 m if the admissible lining infiltration is reduced to only $25 \, l/m^2$ per day, provided it complies with the cited DIN standard. Through the same reasoning (a little simplified), it turns out that the immediately larger, commercially available diameter for porous piping, which is 125 mm, is recommended for canals having between 15 and 30 m of transverse section perimeter.

For larger canal sections, it would be necessary to either proportionally increase the piping diameter or, better still, increase the number of drainage pipes, so that sufficient admission surface is obtained between all of them.

As has already been stated, there are many types of commercially available piping that can be employed in drainage systems. The various shapes, dimensions and characteristics vary so much that we will limit ourselves to saying that the designer should guarantee that the water admission capacity per metre length of canal is sufficient.

If the piping material to be employed is porous concrete, it should be required to comply with the corresponding ASTM C-654 standard (6), which refers to porous piping. This requires 0.8 l/min and metre length of pipe for an inner diameter of 10 cm, with a water load of 0.30 m over the lower straight line. It can be assumed that the flow rate entering the pipe is proportional to the exterior pressure measured on the lower straight line of the pipe, so that assuming that in the worst case scenario, the water level soaking the ground merely covers the pipe, it may be taken that the water entering is in the order of 0.25 l/m: Employing piping that complies with this standard will make it easy to achieve the condition in which the water admission is sufficient with respect to the canal infiltration.

The drainage pipes are usually installed in a trench that is filled with filtration material (Figure 9.6). This material is essential when the pipes accept the water through their joints, since the path this will follow will be first through the filtration material and then along through it in parallel with the pipe until the closest joint is reached. Depending on the class of piping, the filter material will require a certain grain size in order to prevent it from entering the pipe and at the same time, avoiding silting up with the surrounding ground. There is a certain type of porous concrete that has a pore size that enables the filtration material to be simply sand, without any special classification being necessary.

However, even in these cases, it may be necessary to employ a special filler material to prevent the silting up of the piping by the surrounding soil.

The position chosen for the drain underneath the bottom does not appear to have any great importance. It would seem the most logical to site it at the centre, following the canal axis (Figure 9.6). This will avoid the possibility of the pipe having to withstand the forces produced by a possible sliding of the slope lining. In spite of this, if the canal is wide and requires more than

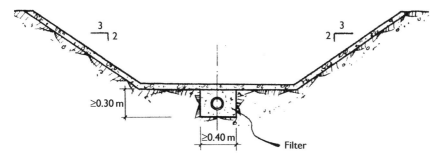

Figure 9.6 Canal with drainage pipe.

one drainage pipe, it would appear that the best plan would be to install one at each side underneath the bottom.

The granular filler material for the trench holding the drainage piping should be correctly compacted, otherwise the bottom might behave as a slab and could fracture.

In addition to sufficient admission capacity for the canal infiltrations by the drainage piping periphery, there are other conditioning factors that require minimum drainage pipe diameters. Normal evacuation of the water transported by the piping is made via a nearby drain or watercourse, which is on a lower level. The evacuation points are thus defined by topographic factors, together with the piping inclination. All this means, therefore, that the piping diameter, the available gradient and existing length from the beginning of the piping to its outlet, must allow the transport of the filtered water flow rate in the corresponding section.

The hydraulic gradient in the drain is formed not only because of the geometric drop between the two ends of the piping, but also due to the admissible pressure under the bottom (Figure 9.7). However, this value is quite small, since it must not be forgotten that an uplift pressure of only 23 cm of water height is sufficient to lift a 10 cm thick concrete lining (because the concrete density is 2.3). Although it is quite true that when the canal is not too wide, the bottom subjected to uplift pressure operates like a beam or slab supported on the lined slopes, which provides a certain complementary strength due to flexion. Nevertheless, it does not seem prudent to exhaust the strength of the concrete with this type of reasoning.

Because of all that has been described, it should not be assumed that at the beginning of the porous piping there is an interior pressure of more than 10 or 20 cm maximum. This therefore means that we have to base our calculations on the fact that the maximum hydraulic gradient is produced by a geometric drop formed by $p + H$, where p, which is ≤ 20 cm, is the internal pressure at the beginning of the porous piping and H is the drop at the porous pipe outlet under the bottom (Figure 9.8). It should not be

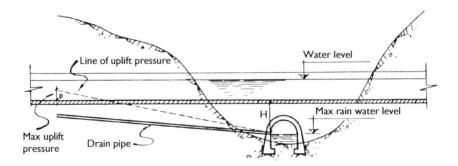

Figure 9.7 Drain gradient.

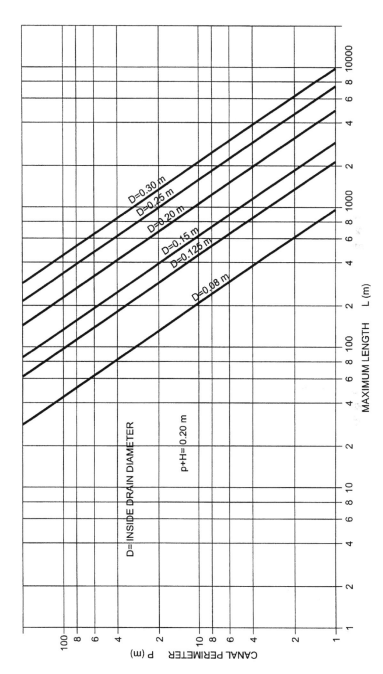

Figure 9.8 For drain length calculation.

necessary to state that the porous pipe outlet should be higher than the maximum rainwater level of the corresponding drain or watercourse.

Strictly speaking, when calculating the maximum hydraulic gradient, the drop involved in the actual canal bottom should be taken into account. However, this slope is usually so small that it can be ignored, although it can be included in the $p + H$ value if so desired.

We will perform the calculation at the limit situation in which the piping is full and is under pressure. If, in this situation, we assume that the diameter is increased, the water circulation could pass to the free level situation, which is more favourable. Therefore, if we perform this calculation assuming a pressurized condition and, in spite of this, are able to evacuate the calculated flow rate, then we will have obtained a sufficient value (although strict) for the drainage pipe diameter.

It should be noted that the filtration contribution inside the piping is made in a continuous approximate fashion, from where the drain begins to its outlet. This means that the flow rate throughout the drainage piping commences at zero and terminates at a maximum value.

It should be remembered that the formulas of Manning, Stricker and Bazin, etc. refer to piping with a constant flow rate along piping being studied, which is not our case on this occasion.

In this type of situation, it is a good idea to calculate the equivalent fictitious flow rate, which is constant along the entire length and produces the same load loss
where:

I is the average hydraulic gradient
l is the generic length from the beginning of the drain to a specific point
L is the total drain length
J is the head loss
Q is the flow rate
K_1 and K_2 are constants.

We have:

$$\frac{dJ}{dl} = K_1 \times Q^2 \tag{I}$$

$$Q = K_2 \times l \tag{II}$$

$$Q_{max} = K_2 \times L \tag{III}$$

From (I) and (II) we obtain:

$$\frac{dJ}{dl} = K_1 \times K_2^2 \times l^2 \tag{IV}$$

where integration produces

$$J_{total} = K_1 \times (K_2)^2 \times L^3/3 \quad \text{or}$$

$$I = \frac{J_{total}}{L} = K_1(K_2)^2 \times L^2/3 = K_1 \times Q_{max}^2/3$$

$$= K_1(0.58 \times Q_{max})^2$$

This means that the average hydraulic gradient is that which would produce a flow rate of $0.58 \times Q_{max}$. In order to simplify a value of $0.5 \times Q_{max}$ is taken, with a certain loss of precision.

Given that L is the length of the drain in metres, from where it begins to its outlet (which is the same as stating approximately that the separation between two watercourses is $2L$) and assuming that the maximum admissible filtration per square metre of canal is 50 l/day, which is equivalent to:

$$\frac{1}{1730} \text{ l/s per m}^2 \text{ is the same as } \frac{1}{1730\,000} \text{ m}^3/\text{s and m}^2$$

and if P is the perimeter of the canal's transverse section in metres and D is the inside diameter of the pipe in metres, and stating the flow rate in cubic metres per second, we can establish the following equation based on Manning's Formula, with a rugosity coefficient of $1/n$ for pipes, which we consider reasonable and set at 60.

$$L \times \frac{P}{1730\,000} \times 0.58 = \frac{\pi \times D^2}{4} \times 60 \times (D/4)^{2/3} \times ((p+H)/L)^{0.5}$$

Figures 9.8–9.11 show tables calculated with this equation in which, for each available $p+H$ drop and for each pipe diameter, the maximum tolerated length for each drain can be obtained without ignoring the fact that it is very easy to calculate the obtained formula with a computer or simple programmable calculator.

Since the choice of drainage points for the longitudinal drain can, at times, be problematic, all engineers have thought, at one time or another, of the possibility of using the actual canal itself as the drain outlet. In fact, there are many canals with a design that follows this idea. The Manimutar Canal in India has been provided with a longitudinal gravel filtration drain that is connected to the canal via holes in the lining, which are 7.5 cm in diameter with a separation of 9 m. This arrangement requires that, before commencing each irrigation operation, these holes have to be plugged with tamped clay, assuming that during the winter, when the canal is empty, these holes are open. Further information with respect to this type of canal solution can be found in the technical paper written by Mr Iqbal Ali, under the title "Some lined canals in South India", which was presented at the "Symposium on Canal Linings", in New Delhi 1960.

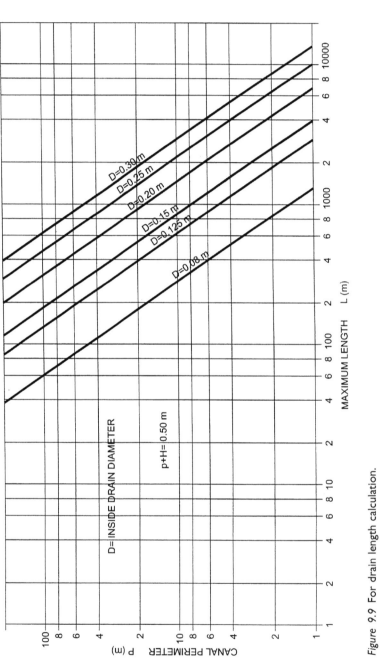

Figure 9.9 For drain length calculation.

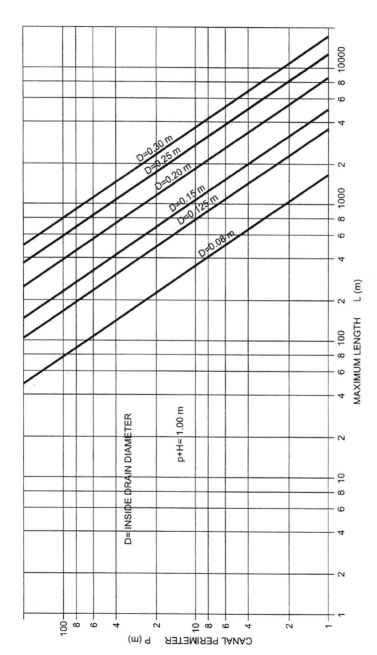

Figure 9.10 For drain length calculation.

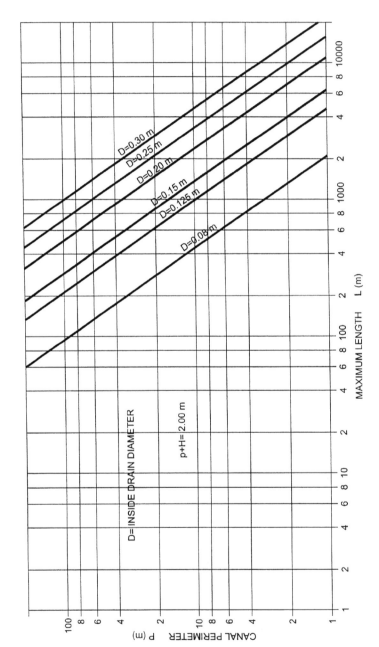

Figure 9.11 For drain length calculation.

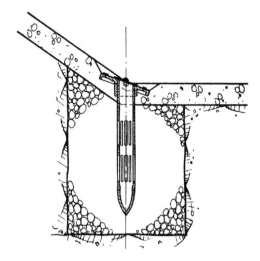

Figure 9.12 Using the canal as drain outlet.

In today's modern lifestyle with its ever-increasingly high wages, this technique does not seem to be very recommendable. Moreover, there is always some doubt as to whether all the holes have been sufficiently plugged, which could otherwise lead to excessive losses, a situation that would be very difficult to correct once the irrigation programme was underway.

Because of all this, engineers in North America have sometime employed this same procedure, but combined with an automatic shut-off valve that will open when the uplift pressure arrives to a certain value (Figure 9.12). The engineer may ask whether these valves are going to jam due to the sediments carried along by the water in the canal, which would prevent them from completely closing, again leading to significant leaks. Quite obviously there remains much to be discussed, however, our own purely personal opinion is that such methods should be avoided in all situations where it is possible to employ drainage piping with an outlet to the exterior. One other significant point is that the latter system permits periodic measurements to be taken to establish whether any canal section requires an inspection of its joints.

9.5   Drainage layer studies

Another method of draining the intrados of a canal lining is to use filtration layers. In general, this makes use of a material that is much more porous than the surrounding ground and which facilitates water circulation through it. Its minimum thickness will be approximately that of the lining.

This type of material is not always easy to install. If it is gravel or ballast with a sufficiently continuous grain size, together with an adequate proportion of fines, such as silt or clay, then compaction is quite easy to carry out, but the permeability is not as great as desired. However, on the other hand, if clean granular material is employed, with more or less uniform diameter, then permeability is greatly improved, but compaction then becomes very difficult.

In order to resolve this problem in an important canal we designed and constructed, we employed a permeable layer of uniform granular material between 5 and 15 mm diameter, to which was added 75 kg of Portland cement per cubic metre of aggregate. The resulting mixture was so difficult to handle that the amount of cement had to be increased to 90 kg/m³. This was applied using the same method as the canal lining, which included a vibro-compaction system.

A drainage layer is an element that possesses characteristics that are basically different to those of a longitudinal drainage pipe. While the latter is able to transport the captured water flow rates, the drainage layer will only permit a very small transport distance since the water circulates with much greater difficulty through the spaces left between the particles of material. On the other hand, however, it facilitates water capture and circulation underneath the lining in any direction.

To a certain extent, the filtration layer is a complement to the longitudinal drain in those cases where it is necessary to encourage the movement of water under the lining towards the actual drain itself (Figure 9.13).

If, for example, the distance between two water outlets in the filtration layer is $2L$, e is the layer thickness and p is the uplift pressure variable existing at the various points along the lining, assuming that this attains a maximum value of po and K is the filtration layer permeability, q is the filtration flow rate per square metre through the lining, l is the variable distance that defines a specific point in the filtration layer, by applying Darcy's Law to the section of layer between two infinitely close points, separated by dl, where the pressure difference is dp, we will obtain the following expression (Figure 9.14).

$$\frac{\mathrm{d}p}{\mathrm{d}l} \cdot K \cdot e = -q \cdot l$$

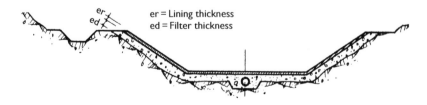

er = Lining thickness
ed = Filter thickness

Figure 9.13 Drainage layer.

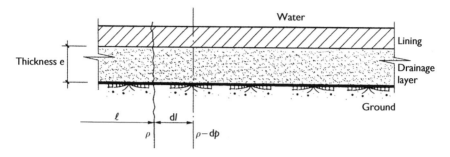

Figure 9.14 Drain layer calculation

The first part represents the flow rate circulating through a layer width of 1 m, with a total section of e. Similarly, the second part is the circulating flow rate, which is equal to that collected in the entire previous section of drainage layer.

This is the characteristic differential equation for the flow through the filtration layer. This produces:

$$K \cdot e \cdot dp = -q \cdot l \cdot dl$$

Integration of this equation provides:

$$K \cdot e(po - p) = q \cdot \frac{l^2}{2}$$

At the outlet and the end of the layer, we have: $l = L$ and $p = 0$, which leads to:

$$po = q \cdot \frac{L^2}{2 \cdot K \cdot e}$$

From which we can obtain $L = (2 \cdot K \cdot e \cdot po/q)^{1/2}$.

If, in this formula, we assume that the filtration layer thickness is 20 cm, the maximum admissible uplift pressure under the bottom is also 20 cm and that the maximum acceptable lining filtration is 25 l/m² during 24 hours, then these values will provide a distance between outlets 2L of 10.70 m with the highly favourable hypothesis that the permeability value K is as large as 10^{-2} cm/s which is equivalent to very permeable terrain.

This clearly shows that the distance covered by the water through the filtration layer is extremely small, but with the penalty that uplift pressures increase in an unacceptable fashion. This means that, in principle, the filtration layer is only valid when complemented by a longitudinal drain which is benefited by the easy arrival of water in the transverse direction.

There is another situation in which the filtration layer could be applicable and this is when it is necessary to protect short backfill sections from infiltration and under which there is a drainage pipe. The filtration layer may be connected to this pipe with only a short run and thus the backfill problem would be eliminated.

Transition zones from excavated areas to backfills are very dangerous in a canal because the excavated areas are normally very well compacted due to the passage of centuries, whereas the transition zone to backfill may undergo settlement, partly due to defective compaction execution and partly because the surface natural ground is uncompressed. Differential settlement in these areas can cause lining to crack, even with well-designed and executed joints, leading to infiltration with dangerous results because the fines may be washed away from the backfill, leading to collapse with subsequent flooding through canal overflow. In addition to taking great care over the quality of execution of these areas, it may also be recommendable to install filtration layers under the lining in order to drain any filtration to the exterior before it has a chance to damage the backfill.

9.6 Analysis of those cases where the employment of internal drainage is necessary in a lined canal

In some canals drainage is installed under the lining and in others it is not. In the former case, after completion of the construction, it is found out that drainage was in fact necessary to prevent damage caused by uplift pressure. Other times, it is installed and the later doubts lead us to ask ourselves whether the associated costs were really necessary.

The most final significant question of all is "when is it necessary to install drainage under the canal lining?"

The answer should be quite clear after all that has been described. If the ground is very permeable gravel or sand, then this will form a magnificent drain and lead the infiltrations from the canal away to lower levels. The lining itself must have a high level of impermeability, since it is exclusively responsible in combating infiltration, but actual drains are not required.

If, on the other hand, the ground is basically clay, then it should be evident that a lining is not necessary to prevent infiltration because if a layer of only 10 cm of clay is sufficient (see Section 2.7), then the ground will be extremely impermeable. If, in fact, a lining is employed, then it will be for other reasons, such as to prevent erosion or the growth of vegetation. In these cases, the lining may be permeable because any water that passes through will be retained by the natural ground and, when the canal is rapidly emptied, will re-enter with the same ease with which it had previously left. The actual canal itself will act as a superb drain rendering drainage piping and materials quite unnecessary.

Another, different situation would occur with clay ground and a sufficiently impermeable lining. Any water that, in spite of everything, managed to escape would be retained between the ground and the lining, resulting in dangerous uplift pressure if the canal is rapidly emptied. In this exceptional situation it would be necessary to install drainage underneath the lining.

The types of ground that require drainage systems under the canal lining are those with a range of intermediate permeability, more or less equivalent to those of silt materials. As we have already seen in Section 2.7, their permeability alone will not guarantee lower losses than the minimum, but, on the other hand, they will not permit easy water circulation through them and so cannot act as drains. In such situations, good linings are required to guarantee low losses, but is also necessary to install drainage systems underneath the lining to prevent dangerous uplift pressures if the canal is rapidly emptied.

Chapter 10

The cross section in a lined canal

10.1 Criteria for obtaining the most economic lined canal

Unlined canals are very demanding with respect to their cross sections, because the shape affects the magnitude of the current's erosive force. In Chapter 11 we shall see that, in weak ground, unlined canals require wide, shallow sections, together with gentle slopes.

The same does not apply to lined canals because their high resistance to erosion makes it possible to practically forget any influence that this phenomenon may have with respect to the cross section, except in specific cases, such as chutes. This obviously means that economics govern the decision concerning the canal's cross section. A similar situation occurs with asphalt concrete linings (except with high speed currents). On the other hand, the problems associated with canals employing a membrane lining with a granular protective covering is the same as unlined canals with respect to their cross sections.

When sufficient technical means are not available for the selection of a determined section for a concrete- or asphalt concrete-lined canal, it would be logical to go for the more economical.

Since the cross area of the canal is the decisive element with regard to the water transport capacity and, on the other hand, the lined perimeter is something that notably increases the costs of the solution. From a purely geometrical point of view, the problem should be tackled by defining the optimum canal section as that which has minimum perimeter for a given cross-sectional area necessary to allow the desired water flow to pass.

The actual geometry itself has resolved the problem for us, showing that the required curve is a half circle. If, as generally occurs, it is not possible to accept a vertical extrados because of slope stability, the necessary solution (also demonstrable) is an arc of circle, the half-angle of which, located at the centre, is the stability limit slope for the ground. This is the form of the cross section of the Zújar Canal, which can be seen in Figure 10.1.

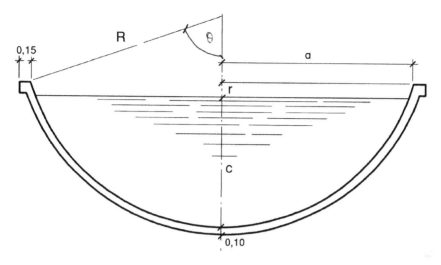

Figure 10.1 Circular cross section of the Zújar Canal (Spain).

The minimum perimeter condition should be found not only to reduce lining costs, which are usually high, but also to diminish friction with the water, optimizing the section still further.

The arc of circle shape for the canal cross section is undoubtedly attractive because it also represents excellent stability with respect to possible slope movement. It should be remembered that the form in which these move is the so-called slip circle, thus demonstrating that the most stable form is precisely that of this type of canal.

If, following the traditional custom, we decide on a trapezoidal canal shape, the geometry will show us, through a simple maximum and minimum calculation making the first derivative equal to zero, that the trapezium enclosing a specific area has a minimum perimeter, which is half a hexagon, in other words a trapezium with base angles of 120° and all sides equal.

From an engineering point of view, this solution is not acceptable because the slopes are too steep to be stable in most types of ground except rock, so that a purely geometric solution is not suitable for obtaining the optimum section. Moreover, this inclination would prevent the construction of linings using slip formwork techniques because the recently poured concrete would be unstable. These slopes are only employed with thick lateral walls that are able to withstand the ground thrust and at the same time provide fixed formwork that is much more expensive.

The most suitable idea would be to obtain the optimum trapezoid shape for a canal, the one with minimum perimeter for a given area, after previously establishing the slope inclination, for basically geotechnical reasons. The solution can also be obtained with simple differential calculus methods,

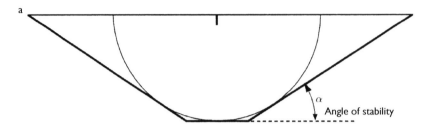

Figure 10.2 Optimal trapezoidal section.

which turns out to be that obtained on drawing tangents, with the same inclination as the slope, to a circle, and extending them to their intersection with the horizontal diameter. In other words, by establishing a trapezoidal section that can be circumscribed around a semicircle (Figure 10.2).

This is a system that provides sections that are very close to the ideal shape. When establishing this, it is also necessary to take other constructional considerations into account, with one of the most important being the machinery to be employed.

When canal excavations are large enough, they are carried out using bulldozers and scrapers, however, smaller canals require the use of backhoes, which results in a higher excavation unit cost.

The bulldozers and scrapers tractor operate in the centre of the canal, excavate the soil, which is then transported to form the lateral backfill, later passing over it to compact it. This requires a minimum canal bottom width of 1.80 m.

The excavators, which are generally fitted with a backhoe blade, or a dragline in the case of ground with water, work at one or both sides as necessary.

If the canal's dimensions do not allow the bulldozer to enter, it might be a good idea to modify the proportions of the trapezium in order to somewhat increase the bottom width and slightly reduce the water height to take advantage of the cost improvement obtained using the bulldozer.

From an economic point of view, the optimum section solution is not obtained through any mathematical calculation, but, in fact, is the result of a series of geometric, geotechnical and constructional considerations.

Everything described in this section basically refers to concrete-lined canals. Another general statement we could make about asphalt concrete-lined canals is that while the water speeds do not exceed 1.5 m/s, they do not present any problems. If this type of lining is used with higher water flow speeds, then the section dimensioning criteria described in Chapter 11 for unlined canals will have to be employed in order to avoid erosion.

10.2 Transitions

Canal cross sections are almost never constant along their lengths. Any changes in slope lead to changes in speed and cross section. In addition to this, the various geotechnical conditions may demand differing cross-sectional shapes, together with flow rate variations, which are frequent in irrigation canals after feeding secondary canals, all of which produces the need to establish transition zones between the various sections.

It is very important to achieve small head losses so that the works can be carried out easily and hence economically.

The simplest form of constructing a transition with concrete between two trapezoidal sections, or where one is rectangular in a specific case, is the result of imagining horizontal straight lines that are simultaneously supported on the two sections of the canal that are to be joined. The figure formed in this way is known as a "hyperbolic paraboloid" in geometry and is quite easy to construct because formwork for concrete is constructed using straight planks, since it is a ruled surface, with no other problems than simply joining them (Figure 10.3).

In order to achieve small head losses, it is recommended that the transition length be a minimum of at least six times the maximum difference between similar widths of both sections, which is the difference between trapezoidal bases or between the upper widths. The head loss formulas were described in Section 1.4.

When the idea is to join a rectangular section with a trapezoidal shape of greater size, which feeds it, cylindrical shaped transitions may be employed,

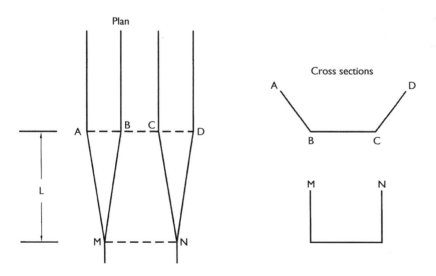

Figure 10.3 Transition of trapezoidal sections.

Figure 10.4 Transition between a trapezoidal section and a smaller rectangular one.

with vertical straight lines, producing inlets and outlets with small losses (Figure 10.4).

If the transition is intended to join a rectangular section to a circular one, the best plan would be to construct flat vertical sides which leave the sides of rectangular section tangential to the circular section and end at the tangential point resulting in triangular figures. The transition is completed with cones, the vertices of which are the union points between the rectangular section bottom and the side walls, and the guideline is the circle of the other section. Figure 10.5 shows one of these transitions in the Castrejón Canal in Spain.

10.3 Freeboard

Freeboard is a depth supplement that is adopted in canals in order to take into account any unexpected events and to provide greater security against overflow.

The causes involved in the adoption of the freeboard widely vary and have very different degrees of importance. In 1954 a congress was held in Algeria, sponsored by the International Commission for Irrigation and Drainage, at which one of the matters covered was precisely the analysis of these problems.

Figure 10.5 Transition between a rectangular section and a circular one (Castrejón Canal, Spain).

The various factors that were examined at this congress are listed below, together with brief explanations:

- With respect to bends, especially where these are pronounced, there is a centrifugal force that tends to excessively raise the exterior level, which can lead to overflow.
- Irrigation water is often loaded with solid sediments, which are deposited on the bottom, reducing the effective section, raising the water level and leading to overflow.
- Aquatic vegetation can be a constant danger in many canals and grows quite readily in the silt deposits on the bottom, which can seriously

aggravate the previously described effect by reducing the useful section of the canal still further.

- There are areas where strong winds exist, which can cause wave action and overflows. This situation can become so extreme in certain areas that the congress recommended that special cross sections are employed for any future canal construction. Small vertical walls are often constructed on the upper slopes in order to reduce the danger of overflow caused by the waves.
- The wave production becomes greater with a longer wind action, which means that the most dangerous sections are those located in the same direction of the strongest prevailing winds.
- Fast gate operation can also cause waves that could lead to overflow.
- Errors may occur during canal operation that can lead to an increase in normal water depth, for example, due to backwater profiles or unexpected flow rates.
- Unplanned flow rates may also become necessary due to unexpected irrigation or population centre supply demands.
- Unexpected heavy rainfall may occur under certain weather conditions and increase the circulating water flow rates if these waters manage to find their way into the canal.
- In addition, all kinds of human error may occur. The designer may commit errors with respect to the canal's rugosity factor, something that is quite probable, because this is something that can worsen with time. The true water heights will therefore be greater than the assumed values, which can cause a reduction of the remnant freeboard and perhaps overflow.
- The constructor can also make mistakes which could result in a milder gradient that is actually desired for more or less long sections. This is quite possible, because the theoretical canal slopes are very gentle, from 10 to 100 cm/km, depending on size. Those sections with a milder slope than the design values may form backwaters, together with the danger of overflow.

All this recommended the adoption of freeboard in order to have a certain margin of safety against overflow, something that is not always easy to calculate, because, in many cases, they do not lend themselves to numeric justification and they also have varying levels of importance according to the existing situation.

In fact, the first of the described causes, which is an increase in normal water depth at bends caused by centrifugal force has enormous significance in small canals and irrigation ditches, which usually follow a layout that is adapted to the edges of properties, leading to closed bends and large angles. In large canals, on the other hand, long straight stretches are the norm, where the layout is not normally restricted to property perimeters, and large-radius bends can be chosen, meaning that this effect is usually negligible and can be ignored; however, all the other factors continue to be important.

This was undoubtedly the reason behind the old 1945 Spanish Ministry of Public Works Instruction, with respect to the Design and Construction of Irrigation Ditches with flow rates of less than 500 l/s, which required that the adoption of freeboard was carried out by only taking into effect the centrifugal force at bends.

The recommended formula was that of Grashoff, which was based on calculating the free water surface at the canal's or river's bends, and which is provided below. This formula can be employed to calculate the difference of water level between the two sides of the same section of canal (see Figure 10.6).
where,

> g is the acceleration due to gravity
> x is the distance from a point on the surface of the water to the bend axis
> z is the water height
> γ is the specific density of the liquid and
> v is the water speed (assumed to be equal at all points).

The resultant of the forces acting on a molecule of water surface at any point must be perpendicular to the final surface curve, otherwise the molecule would move up or down and hence modify it.

The figure shows two right angled triangles. One is formed by the vertical weight force, the horizontal centrifugal force and the hypotenuse is the resultant force.

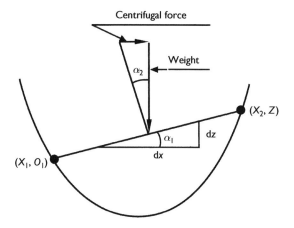

Figure 10.6 Grashoff Formula.

The other is an infinitesimal triangle consisting of dx and dy, which are the distances from the point in question to an infinitely close point.

Both triangles are similar, because they have perpendicular sides, meaning they have equal angles and proportional lengths.

By a property of similar triangles, the quotient of the two sides adjacent to the hypotenuse of one triangle is equal to that quotient of the other:

$$\frac{dz}{dx} = \left(\frac{\gamma}{g}\right)\left(\frac{v^2}{x}\right)\left(\frac{1}{\gamma}\right)$$

where, the first factor of the second member is the mass of water per unit of volume, and the second is the centrifugal acceleration. The final denominator is the weight or stated vertical force.

By integrating this differential equation which is separable, we will have:

$$g \cdot z = v^2 \cdot \log(x) + cte$$

If we refer all the water levels to the lowest point (the internal section of the curve) and we enter the values into the equation, we will obtain:

$$z = \frac{v^2}{g} \cdot \log(x_2/x_1)$$

where x_2 and x_1 are the canal bend radii referred to the exterior and interior banks respectively. This is the equation that will give us the total increase in level for the highest point on the curve with relation to the lowest point.

For the freeboard study, what we are really interested in is not the total water level, but, instead, the maximum increase in level in relation to the average level of the uniform flow. Without too much error, this value is half that obtained by the formula.

Napierian logarithms are employed in the formula. If decimal logarithms are required, as in the first formula from Grashoff, it only has to be remembered that a napierian logarithm is 2.3 times the decimal logarithm.

One practical rule would be to adopt as freeboard for ditches and small canals the increase of level because the centrifugal force is increased by 20%.

For large canals, all the other factors have a decisive influence. Since they cannot be calculated mathematically it is preferred to obtain them through experiment.

In India, it is normal to construct large canals with a freeboard of, some 60 cm.

It is logical that the freeboard is reduced with the canal size, but certain minimum values should be adopted, below which, even in the case of very small irrigation ditches, there is not enough defence for some of the problems to which the freeboard has to face. We have estimated some 15 cm as the minimum value for canals executed onsite.

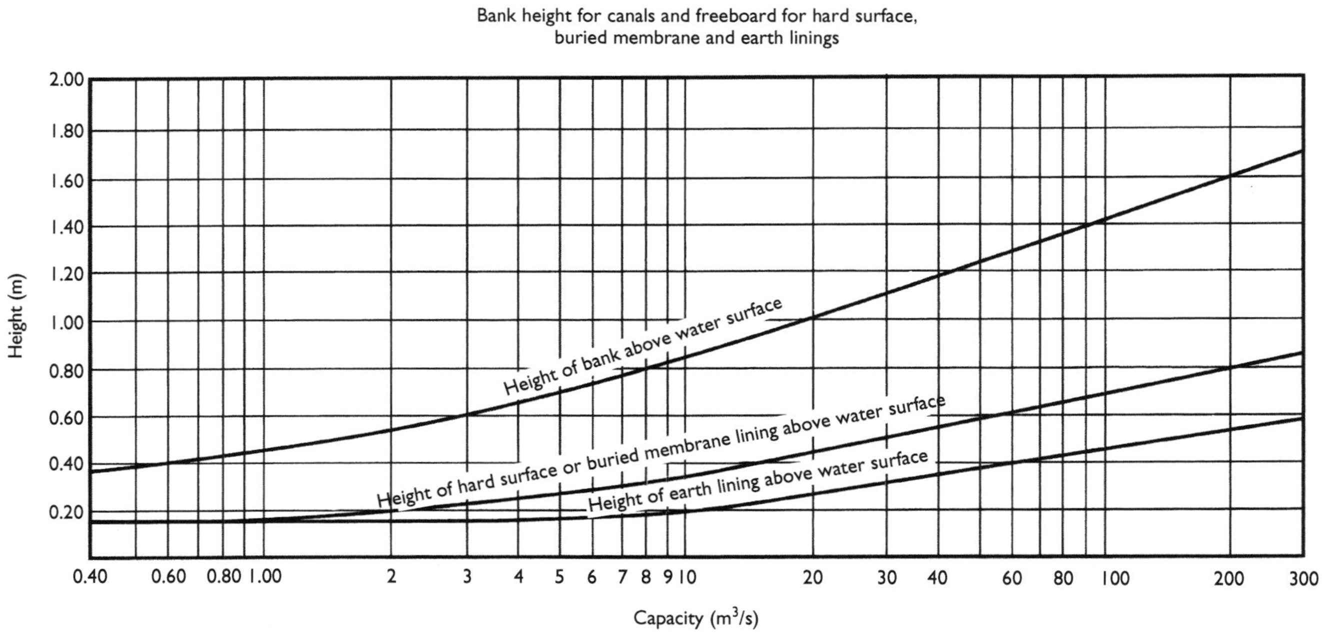
Bank height for canals and freeboard for hard surface,
buried membrane and earth linings
Height (m)
2.00
1.80
1.60
1.40
1.20
1.00
0.80
0.60
0.40
0.20
Height of bank above water surface
Height of hard surface or buried membrane lining above water surface
Height of earth lining above water surface
0.40
0.60
0.80
1.00
2
3
4
5
6
7
8
9
10
20
30
40
60
80
100
200
300
Capacity (m³/s)

Figure 10.7 Freeboard curves.

We have constructed a large number of canals in accordance with all this, with the freeboard equal to one-sixth of the water height, but always maintaining a minimum of 15 cm. For canals having 3.6 m of water height, the freeboard coincides with the Indian experience.

There are certain graphs that are currently accepted internationally, such as that in Figure 10.7, prepared by the Bureau of Reclamation, which provides us with the freeboard to be adopted in canals having various types of linings. In these, a distinction may be made between the actual freeboard itself and that existing from the earth bank or platform found on both sides of the canal, often in the form of a pathway, and which can form additional freeboard in case of need. It can be seen that the freeboard suggested by the graph coincides, in general, with those used by us.

10.4   Covered canal sections

Canals carrying drinking water for the supply of population centres are in danger of being employed for the dumping of contaminating substances that are harmful to health.

Waste dumping into canals, above all in locations that are relatively close to population centres, is a frequent occurrence in spite of suitable measures being taken by the police. Contamination by pesticides and fertilizers used in areas close to the canal, together with contamination by chemical substances from industrial areas, can also be a serious problem.

Although it is quite true that water for human consumption is required to undergo specific treatment to make it perfectly safe for drinking, there are still a few places where this is not yet performed with full guarantee, or where even the very best of treatment is unable to completely remove bad smells or tastes. In any case, public opinion is quite contrary to the sight of seeing how the water they are going to drink is polluted.

This is why it is quite frequent for canals supplying drinking water to population centres to be covered as a measure of preventing this type of pollution.

This covering is a structure for which the costs increase rapidly with the upper width of the canal. It is therefore logical that canals supplying drinking water have cross sections that do not comply with the criteria defined in Section 10.1.

The result is that these covered canals are only rarely trapezoidal in shape in order to avoid large-span covering structures and although they frequently have rectangular sections, other more sophisticated shapes are also possible so that the covering is no longer an addition, but forms a single assembly.

Not only a covering section is sometimes employed to provide protection from contamination if the canal is transporting drinking water, but the

Figure 10.8 Torina Canal (Spain).

canal section may also be covered with soil in hot climates to maintain the water at a cool temperature.

The canal section that is raised above the surrounding terrain, with self-standing lateral walls, whether covered or not, may be an interesting solution in gypsum-bearing soil in order to prevent contact with the ground. The foundations may be constructed using lime paste stonework or sulphate-resistant cement, all of which is covered in Section 12.3.

The side walls are an example of elements that act not only to prevent filtration, but also to withstand interior hydraulic and other forces, in this case, forces produced by the dome.

However, not only canals supplying drinking water are covered. Any canal constructed at the middle of a very steep slope may involve the danger of being choked by alluvion falling from the side. Such is the case shown in Figure 10.8 which shows the Torina Canal (Spain) for hydroelectric use, which has a covered section as a protection element. Canal of Las Dehesas (Extremadura, Spain) is covered at its beginning to provide protection against alluvion from the side (Figure 2.5).

Chapter 11

Unlined canals

11.1 General

There are a large number of unlined canals throughout the world. Their operation is varied, since many of them work perfectly, while in others there are problems with sedimentation and silting or erosion.

These are the most significant inconveniences of this type of canal and this makes them more difficult to design than others, in which the resistant lining is able to withstand possible erosion and where silting problems can be avoided by designing them with sufficient speed. In unlined canals, when sufficient speed is obtained to prevent sedimentation, there is the accompanying danger of causing such levels of erosion at certain spots along the canal that eventually destroy it.

The risk of designing unlined canals that do not last very long is because one of these two problems is much greater than when designing lined canals.

One other point that must be examined with great care is the matter of filtration. Whatever the type of canal, the maximum permitted leakage is 25 or 50 l/m^2 and during 24 hours. This requires ground that is sufficiently impermeable, which in turn requires a prior analysis or verification of the design.

The result is that, because of the described inconveniences, many types of ground are not suitable for the construction of unlined canals and it is therefore necessary to select solutions employing linings.

There are two different theories with respect to the design and calculation of unlined canals. The first is mainly developed in Europe and is called the "Tractive Force Theory". This is based on obtaining an erosive or tractive force of water that is less than that which the ground can withstand, but at the same time is high enough not to produce sedimentation.

The second type is mainly developed in certain British Commonwealth countries, such as Pakistan and India, where it is normally used successfully. It is called the "Regime Theory", referring to the fact that the canal itself in a situation or regime of balance between erosion and sedimentation.

11.2 Tractive force theory

Tractive force, or shear force, was first examined in Chapter 1. We may define it as the erosive force of the water in movement in a canal and its origin can be found in the friction between the water and the material on the bottom. It is also named "shear stress". Its obtained average value is $t = 1000 \cdot R \cdot J$, where R is the hydraulic radius of the water in metres and J is the slope. The tractive force is measured in kg/m^2.

The average value concept comes from the fact that, in this formula, the tractive force has been assumed as being constant over the section's entire perimeter, something that is not, in fact, true.

If we accept the Chèzy formula of $v = c \cdot (R \cdot J)^{1/2}$ as being valid, the average tractive force turns out to be $t = 1000 \cdot (v/c)^2$, from which we can immediately arrive at the intuitive assumption that in order not to erode the canal it is necessary to achieve a reduced speed, until the tractive force of the water is less than that which the ground is able to withstand.

Even at the beginning of the studies into this theory, tables were produced that related the ground categories of the canal with the maximum acceptable water speeds. An example is given below:

Sandy clay

Poorly compacted	0.45 m/s
Average compacted	0.90 m/s
Compacted	1.30 m/s
Highly compacted	1.80 m/s.

The degree of compactness is defined by the coefficient of porosity as follows:

Poorly compacted	Porosity from 1.2 to 2.0
Average compacted	Porosity from 0.6 to 1.2
Compacted	Porosity from 0.3 to 0.6
Highly compacted	Porosity from 0.2 to 0.3.

In granular ground, it may be assumed that the tractive force required to move the grains greatly increases with diameter.

In fact, each particle, assuming a spherical shape, has a weight proportional to the cube of its diameter, and hence the resistance that opposes being dragged along, which is its weight multiplied by friction, is also proportional to the cube of its diameter. Since the force of the water acting on the particle is proportional to the surface area that this presents in the direction of movement and if it is spherical it is proportional to its diameter raised to the second power; it turns out that the tractive force necessary to move a particle increases with the quotient of these values, in other words with the diameter.

The problem is, in fact, complex because turbulences, bottom rugosity and the liquid's viscosity in certain cases, etc. also intervene.

For granular soils, Schoklitsch provides the following acceptable values in function of the grain diameter in millimetres:

$d = 0.005$ mm	$v = 0.15$ m/s
$d = 0.05$ mm	$v = 0.20$ m/s
$d = 0.25$ mm	$v = 0.30$ m/s
$d = 0.50$ mm	$v = 0.40$ m/s
$d = 0.75$ mm	$v = 0.48$ m/s
$d = 1.00$ mm	$v = 0.55$ m/s
$d = 2.5$ mm	$v = 0.65$ m/s
$d = 5$ mm	$v = 0.80$ m/s
$d = 7.5$ mm	$v = 0.90$ m/s
$d = 10$ mm	$v = 1.00$ m/s
$d = 15$ mm	$v = 1.17$ m/s
$d = 20$ mm	$v = 1.30$ m/s
$d = 30$ mm	$v = 1.53$ m/s
$d = 40$ mm	$v = 1.77$ m/s
$d = 50$ mm	$v = 1.97$ m/s

Two tables are provided in this respect, containing values from the USSR and the USA, for both non-cohesive and cohesive soils. These values were published in the Russian technical Magazine "Hydrotechnical Construction" (Moscow) in May 1936 and transformed into graphs, with American units, which appeared in the book of Professor Ven-Te-Chow "Open-Channel Hydraulics" (98).

The permissible velocities for non-cohesive soils, in function of the grain diameter in millimetres are:

$d = 0.005$ mm	$v = 0.18$ m/s
$d = 0.05$ mm	$v = 0.22$ m/s
$d = 0.25$ mm	$v = 0.35$ m/s
$d = 0.50$ mm	$v = 0.42$ m/s
$d = 0.75$ mm	$v = 0.55$ m/s
$d = 1.00$ mm	$v = 0.6$ m/s
$d = 2.5$ mm	$v = 0.70$ m/s
$d = 5$ mm	$v = 0.80$ m/s
$d = 7.5$ mm	$v = 0.95$ m/s
$d = 10$ mm	$v = 1.02$ m/s
$d = 15$ mm	$v = 1.25$ m/s
$d = 20$ mm	$v = 1.50$ m/s
$d = 30$ mm	$v = 1.80$ m/s
$d = 40$ mm	$v = 2.00$ m/s

The permissible velocities for cohesive soils in function of the voids ratio are:

Voids ratio	Sandy clay	Clay
0.2	$v = 2.05\,\text{m/s}$	$v = 1.85\,\text{m/s}$
0.3	$v = 1.75\,\text{m/s}$	$v = 1.45\,\text{m/s}$
0.4	$v = 1.4\,\text{m/s}$	$v = 1.22\,\text{m/s}$
0.5	$v = 1.22\,\text{m/s}$	$v = 1.10\,\text{m/s}$
0.6	$v = 1.12\,\text{m/s}$	$v = 1.0\,\text{m/s}$
0.7	$v = 1.02\,\text{m/s}$	$v = 0.9\,\text{m/s}$
0.8	$v = 0.97\,\text{m/s}$	$v = 0.83\,\text{m/s}$
0.9	$v = 0.90\,\text{m/s}$	$v = 0.78\,\text{m/s}$
1.0	$v = 0.83\,\text{m/s}$	$v = 0.72\,\text{m/s}$
1.5	$v = 0.52\,\text{m/s}$	$v = 0.40\,\text{m/s}$

Unlined canals may be designed using these tables. It is only necessary to select width, water height and slope so that the speed indicated in the table is not attained, together with a safety factor to take into account other causes of erosion, such as curves, which will be dealt with later.

However, it is a better plan to use the tractive force formula $t = 1000 \cdot R \cdot J$, since this explicitly indicates the intervening factors, which are hydraulic radius and slope. In order to reduce the tractive force, it is necessary to reduce the hydraulic radius or the slope or both.

It should be remembered that the hydraulic radius is the relationship between the useful area and the wet perimeter of the canal's section, but in a certain fashion this is also an index of the section's shape (Section 1.3). In a first approximation, it can be said that the section is wider and shallower with a smaller hydraulic radius. This is the reason why unlined canals are frequently wider and have less water height than lined canals, in order to reduce the tractive force.

On the other hand, canals which are deep and narrow, possess greater hydraulic radii and therefore corresponding tractive forces also tend to be greater.

11.3 Improving the tractive force formula

As stated in Chapter 1, the Manning formula is $v = (1/n) \cdot R^{2/3} \cdot J^{1/2}$. This is better than that of Chézy, $V = c \cdot (R \cdot J)^{1/2}$, since it adjusts the coefficient c in function of the hydraulic radius, in other words depending on the section shape.

Both provide a water velocity in metres per second if the hydraulic radius R is given in metres and the gradient J as a decimal value.

In the same way, it is possible to improve the tractive force formula of $t = 1000 \cdot R \cdot J$, by including the section shape in some way through the hydraulic radius.

The Manning formula may be used in the following form:

$$V = (1/n) \cdot (R^{0.16}) \cdot (R \cdot J)^{1/2}$$

By squaring and taking into account that $1000 \cdot R \cdot J$ is the tractive force, we get,

$$V^2 = (1/n^2) \cdot (R^{0.32}) \cdot (t/1000)$$

from which we can obtain

$$t = V^2 \cdot n^2 \cdot 1000/R^{0.32}$$

This means that in two canals possessing the same velocity, in which water height, in other words the first approximation of the hydraulic radius, is different, the tractive force is less in the canal having the greater water height because R is the denominator.

In reference to this, Czech engineers have published some tables containing the maximum acceptable speed in function of the ground characteristics and the water height. These are provided below, and it can be seen that the acceptable speeds are higher with increasing water heights.

Recommended average maximum speed values – For non-cohesive soils *(According to Czech engineers) Recommended speed, in metres per second*

Particle depth size (mm)	Depth in metres						
	0.3 m	0.6 m	1.0 m	1.5 m	2.0 m	2.5 m	3.0 m
0.005	0.12	0.14	0.15	0.16	0.17	0.18	0.19
0.05	0.15	0.18	0.20	0.22	0.23	0.25	0.25
0.25	0.25	0.27	0.30	0.35	0.35	0.35	0.37
1.0	0.45	0.50	0.55	0.60	0.65	0.65	0.70
2.5	0.52	0.60	0.65	0.70	0.75	0.80	0.80
5.0	0.65	0.70	0.180	0.90	0.95	0.195	1.00
10.30	0.80	0.90	1.00	5.50	1.15	1.20	1.25
15.0	0.95	1.18	1.20	5.30	1.40	1.45	1.50
25.0	1.10	1.25	1.40	1.55	5.60	5.70	1.80
40.0	1.45	1.60	5.80	2.00	2.50	2.20	2.30
75.0	1.90	2.20	2.40	2.60	2.80	2.90	3.00
100.0	2.20	2.40	2.70	3.00	3.10	3.20	3.40
150.0	2.60	3.00	3.30	3.60	3.80	4.00	4.50
200.0	3.10	3.50	3.90	41.30	4.50	4.70	4.90

In addition to this one, the Czech engineers have also published a table for cohesive soil, which is shown below:

Maximum recommended of acceptable maximum average values for cohesive soils (*according to Czech engineers*)

Soils	Ratio of pore volume to particle volume	Depth in metres				
		0.3	0.6	1.0	1.5–2.0	2.5–3.0
		Acceptable speed in metres per second				
Sandy clay with	2.0–1.2	0.35	0.40	0.45	0.50	0.55
less than 50% sand	1.2–0.6	0.7	0.8	0.9	1.00	1.10
	0.6–0.3	1.05	1.20	1.30	1.45	1.55
	0.3–0.2	1.45	1.65	1.80	2.00	2.15
Silty clay	2.0–1.2	0.32	0.35	0.40	0.45	0.50
	1.2–0.6	0.67	0.75	0.85	0.95	1.00
	0.6–0.3	1.00	1.15	1.25	1.35	2.05
	0.3–0.2	1.35	1.55	1.70	1.90	2.05
Clay	2.0–1.2	0.27	0.30	0.35	0.40	0.42
	1.2–0.6	0.65	0.70	0.80	0.90	0.95
	0.6–0.3	0.95	1.10	1.20	1.30	1.45
	0.3–0.2	1.30	1.55	1.65	1.80	2.00
Clay without structure	2.0–1.2	0.25	0.30	0.32	0.35	0.38
	1.2–0.6	0.55	0.62	0.70	0.75	0.85
	0.6–0.3	0.85	0.95	1.05	1.15	1.25
	0.3–0.2	1.10	1.25	1.35	1.50	1.60

11.4 Localized erosion: Choice of trapezoidal transverse section

Because the tractive force is not constant along the canal's cross section, the methodology described up to this point can not completely guarantee a lack of erosion since there will be localized points with greater tractive force.

There are two basic sources of dangerous points: the variation in tractive force along the transverse section and the increase in erosive force at bends. In Section 11.6 we will examine the erosive force at bends.

The law of the variation in tractive force in a trapezoid section was studied in Chapter 1. It was seen there that although in trapezoid sections the tractive force on the bottom is 97% of the average tractive force in the section, in the lower part of the slope, before the confluence with the bottom, there is a point with a relative maximum of tractive force, which reaches 75% of $1000 \cdot y \cdot J$, where y is the maximum water height and J is the slope. This point, which is located near to the confluence with the bottom, is especially dangerous because the slope inclination increases the danger of sliding. Experience shows that canal erosion has been preferably produced at this point, as shown by the example described below.

One of the largest canals in Spain is the Castrejón Canal, used for hydro-electric purposes, with flow rates of up to $280 \, \text{m}^3/\text{s}$. Although it has various cross sections, they are all around 7 m water height.

One stretch of the canal was lined with prefabricated slabs, and after a certain length of time, filtration was discovered through the slope. The canal was emptied so that repair work could be carried out and it was found that certain slabs had been lifted exactly at the point we previously indicated as being especially dangerous. The adjoining slabs, located above, had slid downwards, no doubt also influenced by the tractive force which, although not as great at this point, is still significant, and without forgetting the danger of sliding involved in an inclined slope, even when stable. The experience clearly shows the existence of points on the cross section that are exposed to greater danger of erosion.

It can therefore be understood why, when dimensioning canals, average tractive force calculation values are adopted, which are less than those that the ground is theoretically capable of withstanding in order to take into account the supplementary effect of the instability of the lower slope in trapezoid sections.

Figure 11.1 is a classic example, which shows the tractive force components acting on a solid particle supported on a point of the slope, together

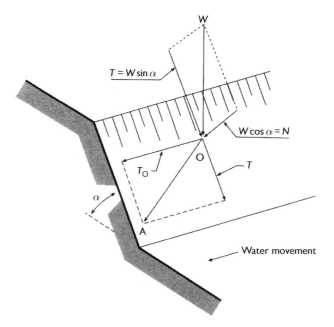

Figure 11.1 Forces acting on a solid particle.

with the particle's vertical weight component over the canal's inclined slope.

where

a is the area that the particle opposes in the direction of the water

w the particle weight

α the slope angle to the horizontal

ϕ the internal friction angle of the ground

τ is the tractive force on the slope

τ_s is the tractive slope on the bottom

y_o is the canal's water depth

y is the water height at any point

The normal component to the slope of weight W is $W \cdot \cos \alpha$

The component parallel to the slope of weight W is $W \cdot \sin \alpha$

The ground friction on the particle is $a \cdot \tau$

The resultant of both drag forces is $(w^2 \cdot \sin^2 \alpha + a^2 \cdot \tau^2)^{0.5}$

The force opposing the drag depends on the normal component and is given by $W \cdot \cos \alpha \cdot tg\phi$.

In order to provide stability, this has to be:

$$W \cdot \cos \alpha \cdot tg\phi = (w^2 \cdot \sin^2 \alpha + a^2 \cdot \tau^2)^{0.5}$$

from which we obtain

$$\tau = \frac{w}{a} \cdot \cos \alpha \cdot tg\phi \cdot \left(1 - \frac{tg^2 \alpha}{tg^2 \phi}\right)^{0.5}$$

Which is the maximum tractive force that a slope can withstand.

At the bottom this will be $\alpha = 0$ and its maximum tractive force will be given by

$$\tau_s = \frac{w}{a} \cdot tg\phi$$

Dividing these two expressions will produce:

$$K = \frac{\tau}{\tau_s} = \cos \alpha \cdot \left(1 - \frac{tg^2 \alpha}{tg^2 \phi}\right)^{0.5}$$

which is the quotient of the maximum tractive forces that can be withstood by the slope and the bottom.

The value of K will always be less than unity, which means that in a trapezoid canal, taking into account the tendency to slide, the maximum tractive force on the slope is always less than on the bottom.

This result may be used to dimension a canal's transverse section, taking into account the problem of greater instability due to slope erosion.

The method of designing an unlined canal with a trapezoid section is as follows:

1   The desired gradient is chosen based on topographic data and canal use.
2   In the cross section the slope inclination is selected in accordance with the ground characteristics.
3   Using the flow rate to be transported as data, the Manning formula is employed with a coefficient of $n = 0.0225$ or 0.0025, together with reasonable values of bottom width and water height.
4   It should then be checked if this tractive force of $1000 \cdot y \cdot J$, multiplied by K is acceptable or not for the ground (y = maximum water height and J = slope).
5   If the ground resistance is more than enough, the gradient can be increased if it is required, or the water height can be increased, and then a return is made to point 3.
6   If, on the other hand, the ground resistance is insufficient, it will be necessary to reduce the gradient and/or reduce the water height, which will involve widening the canal and returning to point 3 once again.

11.5 Optimum section for an unlined canal

In Section 10.1, we examined the most suitable shape for concrete- or asphalt concrete-lined canals, together with those with exposed plastic membranes if they are of very good quality and very careful construction. These do not normally take into account any effects of erosion on the lining because the water speed does not usually reach dangerous levels. However, it must be emphasized that in canals with membrane linings, it will be necessary to employ extreme precautions in the case of relatively high water speeds.

As we have already seen, the case of unlined canals is very different, with great significance placed on the erosion and sedimentation problems.

For this reason, we have explained the methodology for dimensioning a trapezoid section canal, with very careful consideration given to the erosive forces.

A better solution would be to construct a canal without a trapezoid section, but instead with an iso-resistant section, one with equal stability against the combination of tractive force and danger of slope sliding, at all points, which is evidently much better than the trapezoid shape with some points that are weaker than others.

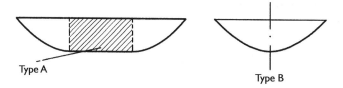

Figure 11.2 Cross sections of equal resistance at all their points.

We have already stated in Chapter 1 that it is a cosine formula and we shall demonstrate this as follows:

α is the angle with the horizontal of the tangent to the required curve at a generic point, with a water height of y. The tractive force at this point per unit of perimeter taking into account its inclination is $w \cdot y \cdot J \cdot \cos \alpha$, and because of the previous is equal to $w \cdot y_o \cdot J \cdot K$. By equalling both values, and substituting K by the obtained value and substituting also $\text{arctag}\left(\frac{dy}{dx}\right)$ by α, gives

$$\left(\frac{dy}{dx}\right)^2 + \left(\frac{y}{y_o}\right)^2 \cdot tg^2\phi = tg^2\phi$$

which is the differential equation for the desired cross section curve for the canal. Integration gives

$$y = y_o \cdot \cos\left(tg\phi \cdot \frac{x}{y_o}\right)$$

which is the equation that defines the section of equal resistance at all its points, taking into account the effect of their variable inclination. A second integration gives the area:

$$A = 2y_o^2/tg\phi$$

As shown quite clearly by Henderson (49), this section is not the only one possible because the result of adding a section of horizontal bottom in the centre part will also meet the condition of iso-resistance since all of it will have the same tractive force as the lower vertex of the obtained curve, as can be seen in Figure 11.2.

The operations involved in the design of an unlined canal with an iso-resistant section are as follows:

1   The desired gradient is chosen for topographic and canal use reasons.
2   A water height is selected, which produces a tractive force that can be supported by the ground.

3　With respect to the previously given section area formula, we will see that the obtained area has either a sufficient value or is insufficient and requires the section to be supplemented with a central part of the flat bottom. If the area is too big, the cross section will be reduced by lowering the water height.

4　The Manning formula is employed to check the hydraulic operation, modifying the water height if necessary, but always reducing it because if it is increased then excessive tractive force will result. The central rectangular part will, of course, also be changed in order to obtain the required transport capacity.

If a solution is not obtained after all these attempts, it may be necessary to go back to point 1 and modify the gradient.

Nevertheless, we must still take into account any possible erosion produced at the bends, which will require that the canal is dimensioned with extra erosion resistance in order to combat this new effect.

11.6　Erosion at bends

Everybody has, at one time or another, stirred sugar in a glass of water with a spoon. It is quite curious to note that any undissolved sugar does not remain uniformly distributed at the bottom of the glass, instead it is concentrated at the centre of the bottom (Figure 11.3).

The explanation can be found in the so-called "limit layer". Due to centrifugal force, which is a consequence of the rotation movement caused by stirring, the water's edges are raised, with a balance existing between the weight of the molecules, the pressure and the centrifugal force, with a form of free level being adopted so that at each point, the result of the forces is normal to the surface, just as was described for the study of freeboard.

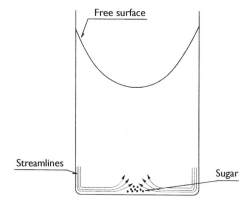

Figure 11.3 Stirred sugar in a glass of water.

However, the water particles in contact with the glass wall encounter friction that reduces their rotational speed and hence, their centrifugal force. These particles, which form the limit layer, are not in balance because the greater hydrostatic pressure at the edges is not compensated by the centrifugal force. The result is that all of them descend along the wall, which establishes a transversal movement as shown in the figure, which forces an interior circulation of water in order to replace the space left free.

If the glass walls were erodible, the water would undercut them, but since they are not, the only thing that happens is that the sugar is transported across the bottom to the centre, and here the water commences an ascending movement and tries to carry the sugar with it, but cannot do this because it is prevented by its weight.

Something very similar occurs at bends in rivers and canals. Also at these locations, in accordance with Grashoff's formula, which we examined during our study of freeboard, the exterior water's edge is raised. Once again the limit layer is unbalanced and a transversal circulation is established, which tends to erode the external margin and transports the eroded materials to the internal part of the bend (Figure 11.4).

The result is very well known: in rivers, the external part is undercut and sediment is deposited at the internal part. The meanders creation begins.

In fact, the phenomenon is slightly modified by the effect of the longitudinal speed of the water, which combines with the transversal to form a movement that is similar to a corkscrew, transporting and depositing the eroded materials somewhat further downstream.

Consequently, it can be said that there is a supplementary erosion effect on the external part of canal bends, with a tendency to deposit sediment at the internal part.

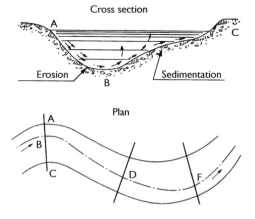

Figure 11.4 Transversal circulation and erosion at bends.

It is not easy to predict the numerical value of this tractive force, which must be added to the previous. What can be said however is that it greatly increases with a reduction of bend radius, and that a certain margin of resistance in the ground must be taken into account in order to defend against this supplementary effect.

11.7 Solid flow

Water not only has the ability to erode the canal perimeter because of its tractive force, but also due to its capacity to transport solids in suspension or those dragged along the bottom produces the so-called "solid flow", which is measured in kilograms of dry material per second.

This material may be the same that the tractive force has stripped away from the canal soil or may be solids that have entered by the main intake with the feedwater.

There are several formulas to measure solid flow, of which the following are just some:

Schoklitsh: $g = 2500 \cdot J^{3/2} \cdot (Q - 0.6 {}^* B \cdot d^{3/2} / J^{7/6})$

where,

g is the solid flow in kg/s
J is the gradient
Q is the liquid flow in m³/s
B is the width in m.

The Meyer–Peter formula is

$g = c \cdot J^{3/2} \cdot (q^{2/3} - q_0^{2/3})^{3/2}$

where,

g is the solid flow in kg/s and metre of width
J is the gradient
q is the liquid flow in m³/s and metre of width
q_0 is the solid flow corresponding to the liquid flow that can begin to transport solid flow as determined by:

$q_0 = 0.6 \cdot (d^{3/2} / J^{7/6})$

Coefficient c depends on the type of material.

The design of an unlined canal using the Tractive Force Theory is carried out by selecting a gradient and a hydraulic ratio that will produce a tractive force of less than that which can be supported by the ground and at the same time capable of transporting a solid flow equal to or greater than that entering the main intake.

If the ground resistance is high, a series of slope values may be encountered for a given hydraulic radius, which are lower than the limit at which erosion commences and greater than the value required for solid flow transport. In this situation, canal design is possible, and even has a range of valid gradients. But if the ground resistance is low, it may be impossible because the slope of no erosion may be less than that required to transport the solid flow.

In such a case, an attempt can be made to find a solution of no erosion by reducing the water height and increasing the cross section width in order to have a more favourable hydraulic radius. However, this will increase the sedimentation problem and then there is no other solution apart from attempting to reduce the transported solids that enter with the water at the canal's intake, so that there is no sedimentation with the low level of tractive force required for no erosion.

This is accomplished using structures called "settling basins".

11.8 Settling basins

Settling basins are civil works installations that remove part of the solid flow through sedimentation of part of the solid flow entering through the canal intake.

First of all, the canal intake must always be located at the external margin of a river bend because here, due to the transverse erosive force effect described in Section 11.6, there is a tendency for the water to carry suspended solids to the other margin, thus avoiding the entry of large material.

Secondly, it should be pointed out that the settling basins (or tanks) operate by achieving very low speeds inside, which causes the sedimentation of the suspended particles at the bottom of this tank, which then involves the additional problem of eliminating this solid material.

In general, the idea of eliminating this material from the bottom of the tank using draglines or clamshell excavators can be rejected because of the economic costs involved, except when this forms good quality aggregate for construction. It can then be loaded onto lorries, assuming that there are consumption locations within reasonable distances.

Under normal circumstances, thought should be given to self-cleaning systems, based on a drain at the bottom which, when opened, produces a strong flow that carries along all the accumulated sediment.

The only place where this can be dumped is in the actual river itself, at a location at a lower level than the tank.

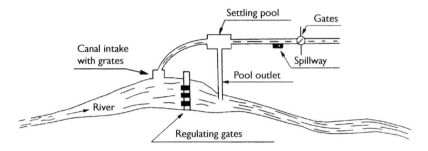

Figure 11.5 Gate arrangement in front of a settling basin system.

Since, in most cases, the river gradient is small, it will be necessary to create a difference in levels, producing backwaters with interleaved gates. Figure 11.5 shows gate arrangement in front of a settling basin system.

In this way, the sedimentation tank can be located next to the backwater part of the river, which will permit the emptying operation to be carried out to the lower level part.

This solid material will therefore return to the same river it originally came out of.

It is recommended that the gates can be raised so that, when necessary, the solid material sedimented upstream by the backwater effect can be allowed to be taken downstream by periodic gate raising operations.

A grille should be installed at the water inlet to prevent the entry of branches and other floating material. Figure 11.6 shows the inlet to the Huaral Canal in Peru.

The tank dimensioning may be performed employing the Dominy–Vetter formula. This formula is given below, together with an example.

W_o is the weight of sediment per cubic metre of liquid entering the canal

W is the generic weight of sediment per cubic metre of liquid that passes a certain transverse section of the settling basin located at a distance of "x" from the beginning

K is the fall speed of the particles that sediment in metres per second

Q is the total liquid flow rate in cubic metres per second entering the settling basin

V is the water velocity in metres per second inside the settling basin

H is the total water height inside the settling basin

B is the desander width, which we assume has a rectangular plan

q is the liquid flow rate in cubic metres per second per metre width of the settling basin.

Figure 11.6 Inlet to the Huaral Canal (Peru).

The Dominy–Vetter formula assumes that at any given moment, the distribution of solid material in the liquid mass is uniform, without taking into account the height of the considered point above the bottom. This may be the case when the solid particles are very small, approximately the case in intermediate sizes and quite false with large sizes. This obviously means that the formula is only exactly or approximately applicable with small sizes of material. However, it must not be forgotten, as has already been stated, that the inlet must be located at a point where large suspended solids do not enter the canal.

In an infinitesimal time of dt, a solid particle will descend Kdt.

In this same time, its horizontal movement, due to being carried along by the water, will be $dx = V \cdot dt$. The descent of this particle will therefore be $K \cdot dx/V$.

The total weight of particles descending throughput the entire tank will be the total W of this section, multiplied by the ratio between the descended height and the total water height, in other words:

$$dW = -\frac{W}{H} \cdot K \cdot \frac{dx}{V}$$

by separating the variables and integrating, we will obtain:

$$\log(W/W_o) = -K/qx$$

with $x = L$ being the total length of the settler and replacing $q = Q/B$ will result in:

$$W = W_o \cdot e^{-K \cdot L \cdot B/Q}$$

This is the formula that will give us the weight of soild material leaving the sedimentation tank, with the amount $W_o - W$ having collected at the bottom.

We would like to point out that the value of soild material that leaves will decrease with the product $L \cdot B$, in other words, with the surface area of the settling basin. This is quite logical, because a large B is required to reduce the speed and a large L to increase the retention time and thus increase the number of sedimented particles.

With the desired value of W fixed, which we can allow into the canal, and knowing the value of K, we can find the surface area LB of the settling tank.

However, both values L and B have to maintain a suitable hydraulic proportion so that the lines of current really occupy the entire settling basin area, avoiding the possibility of a situation of a wide and short plan, in which full advantage of the tank is not obtained. The relationship of $L > 6 \cdot B$ must be obtained.

This formula is employed as follows:

It is necessary to know the flow rate of transported solids W_o, which requires a sample of the liquid entering the settling tank from the river and to weigh the amount of dry sediment deposited per cubic metre of liquid and using this to establish the equivalent average diameter by sieving, for example, through $d50$.

The descent speed K of the solids in the water may be calculated by several formulas, among which can be found the following one:

$$K = ((s-1)4/3 \cdot g \cdot d/c)^{1/2}$$

where,

    d is the diameter in centimetres of the material
    s is the relative density of the material
    g is the acceleration of gravity, $980 \, cm/s^2$
    c is the viscosity coefficient from the Reynolds Number which, for practical purposes, can be taken as being 0.5.

The amount of sediment leaving the tank W, measured in kilograms per second, must be equal to that transported by the canal, in other words, calculated by one of the formulas in the previous section. Figure 11.7 shows the settling tank for the Huaral Canal in Peru.

Figure 11.7 Settling tank for the Huaral Canal (Peru).

11.9 Concept of the Regime Theory

The Tractive Force Theory is based on the attempt to obtain a stable canal by avoiding canal soil erosion through achieving a tractive force that is less than the ground's resistance. The complement of this theory is to achieve a lack of sedimentation until a canal is obtained with no erosion and no sedimentation.

The Regime Theory is based on a completely different concept. The idea is to find a canal in a balanced condition, hence the name, by establishing a situation of in which erosion and sedimentation balance each other over time.

The procedure came into being when the English Engineer Kennedy made a study of canals in India that were in balance and empirically produced a formula that now bears his name and which established the condition to relate speed and water height so that the canal is then in this balanced situation.

The formula is:

$$V = 0.84 \cdot y^{0.57}$$

This formula comes in the old Imperial units of speed in feet per second and water height in feet.

If the true speed and water height of the canal, for example, found using Manning's formula, comply with Kennedy's formula, then it is valid to believe that the canal is in this balanced situation. If the true speed is higher, then an erosion situation exists, and if it is less, then sedimentation is occurring.

This idea met with strong opposition in Europe, where it was felt that this might be valid for certain types of ground found in India, but its validity was very doubtful in other countries, especially taking into account the fact that it lacked any justifying theoretical study to back it up. It was also established that the formula's coefficient and the exponent could require a certain adjustment to adapt them to other type of ground.

It was the Bureau of Reclamation, the important irrigation body in the United States, which established that the formula could be valid, with slight modification, with a water height exponent of 0.5 instead of 0.57, and a range of values for the c coefficient in function of the terrain type.

The result is the graph shown in Figure 11.8, which is expressed in metric units. This can be employed to design canals by selecting geometric specifications so that the speed coincides with that given by the graph for the indicated type of ground.

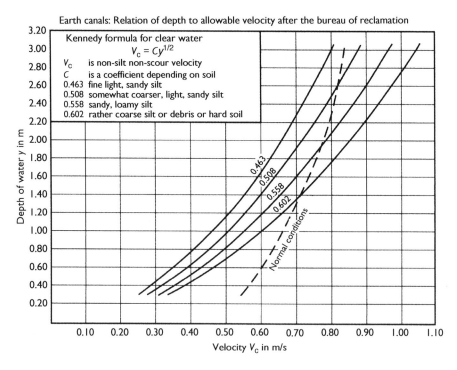

Figure 11.8 Kennedy formula, modified by the Bureau of Reclamation.

However, no mathematical proof for this formula has ever been found, in spite of a great deal of effort having been put into achieving this.

It was Lacey who continued to investigate Kennedy's theory and this, together with the work of other specialists, resulted in not one, but five equations, which can completely define a canal in function of its flow rate and the nature of the surrounding ground.

The five equations are as follows:

$$f = 1.76(d50)^{1/2}$$

$$A = 1.26 \cdot Q^{(5/6)}/f^{(1/3)}$$

$$S = f^{(5/3)}/(1.839 \cdot Q^{(1/6)})$$

$$Pw = 8/3 \cdot (Q)^{1/2}$$

$$V = 0.7937 \cdot Q^{(1/6)} \cdot f^{(1/3)}$$

All these are given in the old Imperial units, in other words, feet, feet per second and cubic feet per second, except for the diameter $d50$, which refers to the average grain diameter and is the only dimension in millimetres. These equations are those now employed for the canals project of the Central Water and Power Commission in India, in accordance with the International Commission for Irrigation and Drainage.

This group of five equations is not yet fully demonstrated from a mathematical point of view; nevertheless, many canals are currently being successfully designed and constructed using them, mainly in India (55), and so this procedure should not be ignored.

This theory is probably most recommended for use with canals in fine grain types of terrain and where the feedwater from the inlet has a solid material load that is to be conserved along the entire length of the canal in order to maintain a stable erosion/sedimentation balance.

It should be pointed out that the area, the perimeter and the gradient of the canal depend on the flow, so that if the latter were to change, then the theoretical canal dimensions must also change, or, which amounts to the same thing, the canal would no longer be in balance. It is therefore desirable that the canal should not vary very much in flow rate, as occurs in many flowing water canals that are fed directly from important rivers and not from a reservoir with regulated flow rates that delivers water according to requirements. If, in spite of everything, the flow rate does vary somewhat over time, then, in our opinion, the theory should be applied using a typical flow rate, which would produce the same result as the range of real rates. This is known as the effective flow rate, which is difficult to obtain, and is the result of various trial calculations.

It should also be pointed out that once the canal flow rate and terrain type have been established, the gradient to be selected is unique. This is another significant difference when compared with the Tractive Force Theory, in which the engineer is able to select the slope from between two extreme values: that which produces erosion and the one producing sedimentation.

11.10 Ground quality for the construction of unlined canals

There are three basic aspects involved in the examination of the ground where an unlined canal may be constructed.

The first is the ground impermeability. It was seen when discussing filtration in Section 2.7, that a Darcy Permeability Coefficient of $k = 2.5 \times 10^{-4}$ cm/s was sufficient to guarantee maximum losses of 25 or 50 l/m^2 in 24 hours. This means that many silts and clays will provide sufficient impermeability.

The second is the tractive force resistance. This specification cannot be analysed in isolation, but instead, by examining it with the forces developed by the canal water.

The third is the cross slopes stability against possible landslides. This specification is common to all canals.

The various types of ground may be examined in the light of these specifications and hence their suitability with respect to the construction of unlined canals may be judged.

We shall make use of the Casagrande nomenclature in order to define the various types of ground. According to this method, each type is associated with one or more letters in accordance with the following rules:

G – gravel
S – sand
M – silt
C – clay
O – organic soil
W – with granular material indicates well graduated
P – with granular material indicates poorly graduated
H – high plasticity
L – low plasticity.

Two letters may be used together to indicate a mixture. Two groups of two letters may also be employed to describe a mixture, in which each group consists of a letter indicating the material and the other indicating a complementary definition.

For example, SP-SC would indicate a poorly graduated sand mixture with sandy clay. It should be remembered that a poorly graduated aggregate means that it does not possess a well-distributed grain size, with either an excess or lack of certain sizes, which leads to poor compactness.

The first material to study is GW-GC (well-graduated gravel with clay). Well-graduated gravel has a level of resistance to erosion that comes not only from the large size of the aggregates, but also from the fact that with a good grain size range, the individual grain fit together well, which provides a strong structure. The mixture with clay increases cohesion, which augments strength still more and also provides impermeability and this makes them the best types of ground for unlined canals.

If the gravel is poorly graduated – in other words, it has proportions of the various sizes that are not adequate with respect to providing good levels of compactness, but in addition contains clay (classification GP-GC) – then this will also form a very good type of ground for canal construction. Although the resistance to erosion is somewhat less than the previous case, it is still very good due to the large size and cohesion contributed by the clay. This provides impermeability, and enables this type to occupy second place with respect to suitability for the construction of unlined canals.

An SW-SC type of ground, with well-graduated sand, and clay, possesses less resistance to erosion than the previous types because the maximum aggregate size is also much less. It should be remembered that, according to the studies by Shield, the grain stability compared to the water speed increases with its diameter in a first approximation. Its erosion resistance is due to the well-graduated sand, which produces a close-fitting structure, together with the cohesion provided by the clay, which continues to be the material providing the impermeability. This type of ground can be described as class 3, with regard to serving as a base for canal construction.

If the ground consists of poorly graduated sand with clay, in other words, is type SP-SC, it is now beginning to involve serious erosion problems because the small size aggregate does not have the support provided by a range of well-fitting sizes, which would give a good structure. Erosion resistance is basically obtained by the cohesion contributed by the clay, which still provides the impermeability. With respect to unlined canal construction, this type of ground would be classed as 4.

Fifth place on the unlined canal ground classification list is occupied by type CL, which is low plasticity clay. Unlike high plasticity clays, which cause problems in all engineering works because they are highly sensitive to changes in moisture levels which can lead to dramatic modifications of volume and thrust, low plasticity clay are much more stable and many canals have been successfully constructed in this type of ground. The Bureau of Reclamation curves, the tables of the Czech engineers and the curves of Kennedy, modified in the USA, all provide data for the construction of canals in cohesive ground. These types of ground are completely impermeable, but require careful attention in the selection of slopes, because they can suffer from sliding.

Lastly, number 6 on the list is GM, gravel with silt. This type of ground lies at the very limit for the construction of unlined canals. Erosion resistance depends exclusively on the large aggregate size, which will improve with graduation or diameter range. Unlike clay, silt possesses no cohesion, and therefore does not contribute anything to stability, or against erosion or slope sliding. In addition to this, we saw in Chapter 2, Section 7, that silt was also at the very limit for guaranteeing filtration losses of around 25–50 l/m² and in 24 hours. Not all silt is of the same type, but in general they provide the accepted limit of impermeability. If the situation is ground with improved soil, in which a canal is to be constructed with a GM lining (Section 11.11), it will be necessary to take into account that the lining thickness, even when it is thick, it may not have enough impermeability because the situation is worse than that studied in Section 2.7.

We can sum up by stating that, in general, unlined canals can be constructed in this type of ground, but they require a very careful, detailed study with a very strict works control.

All other types of ground are not suitable for unlined canal construction, unless very significant corrective measures are undertaken. Some of these will present special problems, which will be discussed in Chapter 12.

When examining some of the unsuitable types, we shall first look at the SM type, sand with silt. As has already been stated, we will be at the very limit of impermeability, but there will not be sufficient resistance to erosion, because there is neither cohesion nor sufficient aggregate diameter.

The ML soils which contain only silt, although they possess low plasticity, are impermeable, but do not have sufficient erosion resistance.

Type CM, which is silt mixed with clay, is impermeable, but it has very little erosion resistance. It cannot be recommended for canal construction.

Types GW, GP, SW and SP are not impermeable and hence are not suitable, independently of some having higher resistance to erosion than others, which will depend on the grain size and range.

Organic soils, whether high or low plasticity, OH or OL are never recommendable for civil engineering and even less so for canals. They are plastic with very little stability, and do not provide any reliability whatsoever.

11.11 Canals in which the ground is improved

At times, the ground in which a canal has to be constructed is of very poor quality, due to either permeability problems or a lack of resistance to erosion. If, in these cases, there is soil available that is of sufficient quality, at a distance for which transport is economic, then a type of lining can be made with this better soil, which is known as "improved soil" (see Figure 11.9).

From a hydraulic point of view, these canals behave like unlined canals and for this reason they are included in this chapter, but as if they were constructed in ground of better quality than the original. Many engineers say that these canals are with thick compacted earth linings.

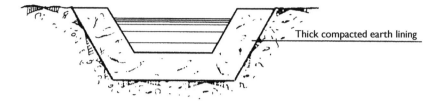

Figure 11.9 Thick compacted earth lining.

The minimum recommended lining thicknesses, (7), (10), are provided below in metres:

Water height	Horizontal lateral wall width	Bottom thickness
Less than 0.60	0.90	0.30
From 0.6 to 1.20	1.20	0.40
From 1.2 to 1.80	1.60	0.60
Greater than 1.80	2.40	0.60

The relatively large thicknesses are quite obvious, above all on the slopes. Water erosion and aquatic vegetation clearance, which has to be carried out periodically, may reduce this thickness and hence, has to be large. All this has been verified by the more than two centuries of operation of the Imperial Canal of Aragon that was constructed on gravel terrain, with an impermeable clay lining, and which has been disappearing in successive peelings.

Also noteworthy is the greater lining thickness on the slopes than on the bottom. This is due to the compacting of material carried out in horizontal layers, which facilitates bottom impermeability, since the water will have to penetrate these layers at 90° to them. On the other hand, the water can horizontally cross the joints between slope layers, a path that is easier and has to be compensated by means of a longer run.

The slopes should be lined with a greater thickness than that actually required in order to profile the excess, which is usually the most poorly compacted because it is difficult for the tractor to reach the edge of the canal.

If the sides of brought-in soil have thicknesses of less than 1.80 m, it will be difficult to obtain compaction machinery capable of providing good results. It will therefore be necessary to examine the possibility of filling the entire space occupied by the canal with improved soil and then profile out the part to be occupied by the water.

Improved soil linings can resolve problems involved in certain unlined canals. If, for example, the ground is type GW but does not contain any clay, then we know that the problem is impermeability, but not resistance to erosion. If good quality clay can be found at a certain distance, it might be possible to transport this to the canal's location. Then the gravel ground can be scarified, the two materials mixed, wetted with an amount of water in accordance with Proctor and finally compacted to obtain the required density, forming the previously described bottom and slopes.

The amount of clay to be mixed should only be fixed after performing adequate trials, at least those referring to gravel graduation. However, the actual amount to be mixed, if the gravel is truly well graduated and contains sand in suitable sizes and proportions, is small, around 10%. It depends on the amount of existing voids.

Something similar may be said of SW types of ground, which can also be improved by the addition of clay, although the low erosion resistance, due to the small aggregate size, requires the matter to be treated with some caution.

11.12 Ground correction: Cement-treated soil, bentonite, chemical gel

The ground where the canal is to be constructed may be improved by other procedures, which are suitable to the actual ground characteristics.

For example, trials have been carried out on cement-treated soil, which have not led to the development of any widely adopted technique for canals, consisting in mixing a certain amount of cement with the surface part of soil that has been previously scarified, together with a suitable amount of water and subsequent compaction in order to produce a lining. Cement-treated soil can also be prepared by taking a certain amount of soil, especially when the grain size range is good, then mixing it with cement and water to form a type of concrete without any large size material, which is then extended and compacted using one of the procedures described in Chapter 5. There are two different systems in use.

The addition of the cement to the soil produces two effects: it increases the amount of fines since cement has a very fine composition, which increases impermeability, and in addition to this, when the cement sets, it increases the strength, establishing a type of cohesion between the fine grains.

The most suitable soils for this treatment are undoubtedly sands and sand-silts (SW and SM) since these possess very little erosion resistance and unacceptable or limit permeability, which sometimes is sufficient and other times is not.

For economic reasons, the cement-treated soil is used in only thin layers of 10–15 cm and this makes it an almost concrete-like lining.

If we make use of our knowledge concerning contraction joints in concrete linings (Chapter 4), it can be stated that the closeness of contraction joints increases with lower lining strength, with greater binding of the lining with the underlying ground and with smaller lining thicknesses. Except for the last factor which, being the cement-treated soil with a thickness equivalent to the concrete lining, does not worsen the situation, the other two predict heavy cracking because the cement-treated soil strength is much lower than that of normal concrete and its union with the subsoil is total.

We can, therefore, predict widespread cracking in these canals, above all if thickness is reduced so much in order to obtain greater levels of savings. Perhaps excessive filtration will not occur, especially when silts or sand-silts are involved, in other words, those possessing almost the required impermeability, and of course the improvement in filtration levels compared to untreated ground is very great. In addition, there is no doubt that there will be a significant improvement in resistance against erosion.

The two construction systems for cement-treated soil have different characteristics. When the soil is mixed onsite with the cement, the result is not stable except with very gentle slopes having values of around 1/4, which also makes life very easy for the compaction machines. On the other hand, if the cement-treated soil is produced in the same way as normal concrete with concrete mixers and paving machines, then the same 1.5/1 slope as the concrete lining may be employed or a more gentler one, as necessary.

In practice, there are more differences. It is easier to produce thinner layers for example, 4 or 5 cm, of cement-treated soil using the onsite system. The amount of cement employed may also be reduced because the mixture may be less workable. These two circumstances have frequently been led to describe the onsite mixing system as producing worse results when the real cause is the poor quality material.

Studies carried out at research centres (38) on test sections have shown filtration levels of $1.34 \, m^3/s/10\,000\,000 \, m^2$, equivalent to $115 \, l/m^2$ per day, which are quite unacceptable when compared to the values we have been using throughout this book of 25 or $50 \, l/m^2$ per day. These filtration levels are attributed to the numerous hairline cracks that appear. However, in general, it is accepted that the reduction in filtration can be very large in the case of favourable types of ground. The University of Civil Engineers in Santander carried out a series of laboratory trials under the supervision of the subscribing professor, according to which the permeability of silt may be divided by almost 100 when 15% cement is added, but without taking into account any possible danger of cracking.

In 1947, a test section was constructed for the lining of an Austin Project Canal (USA) (38). The SP-SM (sand-silt) ground was scarified, mixed with cement and water and then compacted as if it were concrete. The resulting layer was 7.5 cm thick, and the subsequent filtration was $37 \, l/m^2$ per day, which was perfectly acceptable. However, in those sections where the amount of cement had been reduced, the water erosion was quite noticeable. It is quite evident that with respect to erosion, the cement-treated soil, especially when it has been mixed onsite, appears much more like an unlined canal than one with a concrete lining because there is no gravel and the size graduation is not perfect, etc.

The amount of cement added must be adjusted by means of test mixes, but it can be predicted that it will be required to fill the material cavities. This will lead to amounts between 7 and 16%, which produces dosages of around 150 kg of cement per cubic metre of soil material. With a treatment thickness of 15 cm, some 25 kg/m^2 of surface will be required. The usual amount of water is between 12 and 15%. These values, taken from normal soil percentages, have been verified by practical studies carried out in the production of prefabricated cement-treated soil slabs (82). There are several ASTM standards (4) available that can be consulted, which deal with the stabilization of cement-treated soil for roadways, but which in practice can also be employed in canal construction.

Figure 11.10 shows the Calagua Canal in Uruguay, which was constructed using cement-treated soil. It can be seen that the actual lining is very thin and, in our opinion, not very recommendable.

This soil treatment is, in fact, a soil stabilization operation. In the same way that roadway works have employed other means of stabilization, similar experiments have also been carried out on canals, but we are not aware of any definitive results.

Once the canal has been constructed and put into service, it is possible to use sealant products. They are mixed with the canal water, almost always

Figure 11.10 Calagua Canal (Uruguay). Lining of cement-treated soil.

when the water is unmoving, by means of temporary small dams at regular intervals along the canal. These products are then carried along by the water and as this filters through ground they become trapped between the particles and fill any cavities. This same effect occurs when the canal is transporting turbid water (Chapter 2).

The most tempting material is Bentonite, which is highly expansive clay. However, it has not been fully established whether these products are really effective, because experience has shown that they are usually retained in the most superficial layer of ground without penetrating down to deeper layers. This then forms an erodible surface crust that can become easily deteriorated. Because of this, trials have also been carried out with the canal empty and the surface soil layer scarified, mixing it with a certain amount of Bentonite and wetting it, followed by compaction. We are of the impression that this method can be successfully employed to impermeabilize reservoirs or storage ponds, but its use in canals has not gained popularity because of the involved costs.

Special resins can also be used, which are intended to increase the ionic attraction of the ground particles by the water, which will then augment the thickness of the hygroscopic wrapping around each particle. These are solid dissolved in liquids that are miscible with water and their manufacturers guarantee filtration reductions of 60–90% in sandy ground. These usually have the same inconvenience as Bentonite, in other words, they have to be introduced into the still canal water by means of small dams. Again they only penetrate the upper ground layer, creating an impermeable, but weak crust. It is obviously essential that these products are not harmful to human health, or to any fauna and flora associated with the canal and the water it carries.

These products should not be employed with just any ground, because they require a certain grain size that enables the wrapping that forms around them to close off the empty spaces between them.

Chapter 12

Canals with peculiar problems

12.1 Introduction

Sometimes, the design and construction of works for a canal take place under especially difficult circumstances. Generally, the problem comes from its location in complicated ground structure, among which we can cite expansive clay, gypsum-bearing soils and loess. However, on other occasions the conditioning factors are due to the actual canal operation itself, as in the case in which an existing canal has to be lined, but for service requirements it is not possible to empty it and the lining works have to be carried out with water still circulating.

When the problem is caused by the design of a canal on complicated ground, the adopted solution always includes an attempt to vary its course in order to locate more favourable land, but since this is almost always impossible, it then becomes necessary to isolate the canal itself from the surrounding adverse environment. When dealing with specific problems imposed by operation of the canal it is not possible to provide any hard and fast rules, instead, an attempt must be made to take advantage of whatever favourable circumstances there may be.

12.2 Canal in expansive clay

Expansive clays are characterized by large changes in volume depending in their moisture content, which are also accompanied by great pressure due to swelling (certain montmorillonites, without any external pressure, can increase their volume by more than ten times under wet conditions in the laboratory). The Darcy permeability factor is extremely small, in the order of 10^{-6}–10^{-9} cm/s.

The suitable cross section includes paying particular attention to the impermeability of the lining, in order to free the ground from the corresponding changes in moisture levels, together with a comprehensive study of drainage and the placing of a layer of better quality soil between the canal's water and the expansive clay deposits in order to form a protective cushion.

The weight of this layer of soil, no matter how thick it is, will not, in general, be able to overcome the pressure caused by the possible swelling of the expansive clay, which may be very high. However, on the other hand, it will always assist in guaranteeing better isolation and impermeability. The reason for giving this intermediate layer an appreciable thickness is that it is also susceptible to possible cracking because of changes in moisture levels. We can therefore summarize its purposes as a desire to homogenize the charges and prevent the moisture reaching the expansive clay.

The expansive clay found in ground where canals have been constructed may exceptionally attain potential volume increases of up to 12% and swelling pressures of $1.7 \, kg/cm^2$ (58). This pressure is the equivalent of the weight of a 10 m thickness of soil with a density of 1.7, which shows that containing its thrust by means of a layer of soil on top is not feasible.

Drainage under the lining is usually essential for the reasons given in Section 9.6.

Figure 12.1 shows the cross section of the Subeybaja-Javita Canal in Ecuador, and the drainage layer can be seen under the concrete lining, together with the longitudinal pipe drainage and the thick layer of selected material separating the lining from the natural ground.

In such canals, 1 m thick layers of protective material are located above the natural expansive clay. The experience of the engineers who have had some contact with these works is that on many occasions this thickness should be increased.

Another very adequate solution would be to employ a plastic membrane lining supported on a thick layer of soil and with a layer of gravel over the membrane, instead of the concrete lining. In fact, this is the same solution that was described in Section 8.1, on membranes with protection, but with an increased thickness of the supporting soil. However, in contrast, the resistance to the tractive force would have been less than with a concrete lining.

An almost identical solution to that adopted for the Subeybaja-Javita Canal was employed in the Genil-Cabra Canal in Córdoba, Spain, which was also constructed in expansive clay. The natural clay was excavated, placing a layer of silty sand under the lining. This was around 1 m thick. The concrete lining is 10 cm thick. One notable feature is that the cross section is parabolic, which, as we have already stated elsewhere, possesses certain advantages similar to those of the circular section. The actual concreting was carried out using a Gunther-Zimmermann longitudinal slip form machine (41).

The study carried out using the finite elements method with respect to the stress produced in the concrete lining due to the effect of the expansive clay (76), (84) is very interesting. The result is that with clay of an average expansivity of around 0.5%, dangerous levels of stress are produced of approximately $5 \, kg/cm^2$, the same as in trapezoidal and circular sections.

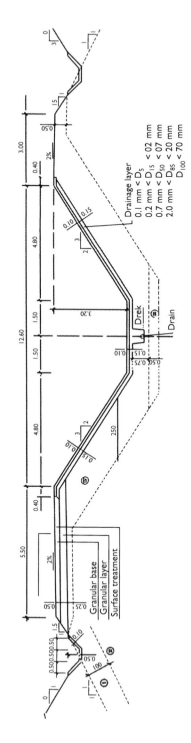

Figure 12.1 Cross section of the Subeybaja-Javita Canal in Ecuador.

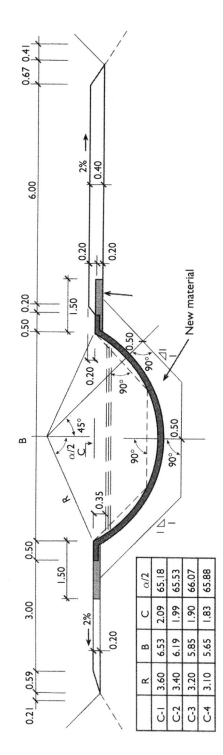

Figure 12.2 Calanda Canal (Spain).

Nevertheless, if a longitudinal joint is located in the lower straight line of a circular section, the tractive forces almost completely disappear and the lining works as an articulated inverted arch. One consequence of this study is the proposal for the Calanda Canal in Teruel, Spain, the cross section of which is shown in Figure 12.2, in which the philosophy coincides with the previous description, but where the main element is a longitudinal joint in the lower straight line.

There are many canals in Spain that have problems with expansive clay, with the Bardenas Canal being worthy of mention.

It is evident that since the clay thrust forces are produced by swelling phenomenon caused by changes in moisture values, it is essential for the joints to be correctly impermeabilized with a suitable, and above all, long-lasting material.

12.3   Canals constructed in gypsum ground

Gypsum is calcium sulphate and is relatively soluble in water, which means soils containing it are in danger of being partly dissolved by the available water and producing internal cavities, which can then lead to sinking of greater or lesser importance, but which invariably causes the destruction of the canal.

To the physical danger of the canal collapsing has to be added that of the sulphates reacting with the free lime in the cement, producing Ettringite, which is extremely expansive and will completely destroy the concrete of which it forms part. The damage produced by gypsum falls into two categories: that of ground failure and lining destruction if it is made of concrete, or even the stonework.

The problem of gypsum-containing ground is widely extended. 25% of the Spanish territory is affected and there are several countries in the Middle East with serious difficulties over most of their area because of this.

The treatment of canals under these circumstances has certain similarities to that for expansive clay. Impermeable linings have to be employed, with drainage that is able to eliminate all possible filtration before it has a chance to reach the ground. However, great care has to be taken to ensure that this same drainage system does not encourage water flow close to the gypsum, since this will dissolve it leading to subsequent collapse.

At times, in Murcia, Spain, a drainage brick layer is installed between the excavated ground and the concrete lining, which conducts the filtrations from any possible cracks to a longitudinal drain. A plastic membrane is also employed to isolate the ground from the drained water.

One solution that is very efficient, but also more expensive, is to raise the canal above the ground, isolating it from the gypsum. It is only necessary to adopt the precaution to construct the foundations with sulphate-resistance

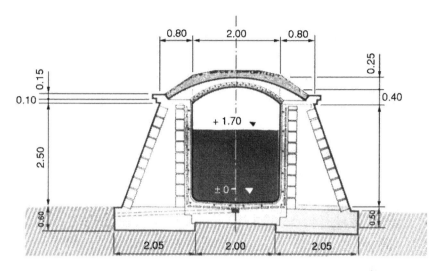

Figure 12.3 Canal above the ground, with foundations in gypsum (Illustration of Canal Bajo for water supply to Madrid).

materials, for example rubble work mixed with lime or special cement (Figure 12.3).

There have also been instances in which the canal or special works are founded over piers. The use of prefabricated units, which are self-standing sections, such as beams supported on bridge piers, may be a magnificent solution in the case of small canals or irrigation canals (see Section 18.3). The foundation of the piers has to be sulphate-resistant. Among the possible types of self-standing units, some have joints between sections just above the supporting pile, while in others, this is located in the middle of the span. Evidently the latter type is more convenient because any possible leaks will not soak the pile foundations and thus this will prevent the gypsum from being dissolved.

It has been observed that the number of fractures produced in canals by gypsum increases when there are watercourses or where it is possible for groundwater to pass underneath the canal. This is caused by the greater hydraulic gradient existing in the watercourse zone producing a larger underground filtration rate with the corresponding dissolution of gypsum. The collapsing produced in the ground leads to the sinking of the backfills located above (53).

In such cases, the danger does not come from the actual water transported by the canal, but instead from rainwater or even groundwater layers fed from distant areas. Here, it is necessary to carry out a study of possible drainage to protect the areas of ground close to the canal that are in contact with these waters.

Exterior drainage consisting of cunettes is very important and should be very carefully examined.

In various concrete-lined canals, the Aragón y Cataluña Canal in Spain for example, enormous amounts of money have been spent in order to inject with clay or clay-cement the cavities produced in gypsum ground, in order to prevent the canal above from sinking. The cement employed must obviously be sulphate resistant. It would be a very interesting exercise to compare its cost over time with that which would have been necessary to protect the canal from this water during the original construction, together with the installation of a more conceptually suitable lining.

Problems with canals in gypsum ground are both frequent and serious in the Ebro basin. In addition to the Aragón y Cataluña Canal, there are similar problems in the de Monegros Canal, the Violada irrigation ditch, together with the Terreu Canal, the Imperial Canal, the Cinca Canal, etc., all of which are located in Spain and also affected to a certain extent.

In relation to the aggressiveness of gypsum-containing ground towards concrete, it should also be pointed out that even the so-called "sulphate-resistant" cements may have problems (14). The reason behind this is quite simple: the sulphates attack the lime freed during the setting process and many special cements resolve this through the chemical reactions produced by other additives having an acidic nature, such as pozzolans, slag and fly ash, etc. (Section 5.7). These reactions require a certain time to become effective and if the concrete is produced and poured onsite, it may be attacked while it is still fresh.

Because of the foregoing, we can state that the use of prefabricated linings, which are installed on site after a certain amount of time has elapsed (Section 6.1), together with the use of prefabricated irrigation canals and other small works is a recommendable scheme for gypsum ground.

12.4   Canals in loess

Loess is a material consisting of very fine silt, with an apparent cohesion that can completely disappear. To this problem must be added the fact that they are not sufficiently compacted and may therefore suffer collapse through a reduction in volume, a phenomenon mainly produced when it becomes wet. These are areas in which the main danger is subsidence.

Geologists affirm that the original source of loess may be found in the very fine sediment produced by glaciers, which were periodically melted in certain geological periods and this sediment was deposited over the plains. When new ice-age conditions returned, this sediment-covered ground was dried out and strong winds carried it off and deposited in more protected locations.

Loess is a very dangerous material and, unfortunately, is abundant in many areas of the world.

There is an abundance of collapsible silt in Spain, which perhaps frequently has a different geological origin to loess, but the very similar characteristics and the associated problems are equally serious. Such materials are very frequent in the Ebro basin.

Collapsible silt consists mainly of around a minimum of 80% of silt, although it can also contain small amounts of fine clay and sand particles. Its main characteristic is that its grains do not occupy the minimum possible volume, but are supported one on the other, producing a great number of cavities between them. On many occasions there are small amounts of crystallized salts which glue some grains to others, leading to an appearance of stability and cohesion which is deceptive and hence extremely dangerous. This cohesion can be completely lost and its granular structure can very brusquely change, either because of external loads or because it becomes wet and this can lead to the serious collapse of any canal that is in contact with it.

The classification of collapsible silt can be made in accordance with the teachings of soil technology, with the instructions provided in the Bureau's *Soil Manual* (31) being very useful.

The applicable treatments are varied and none is definitive. If it can be compacted then we can be certain that this process will be beneficial because it will produce at least a partial consolidation prior to the works execution. The first solution to be examined in the case of difficult terrain should always be that of moving the canal's route. If this is not possible, a solution must be chosen that will isolate the canal from the ground. The usefulness of self-standing canal structures in gypsum ground also applies here, providing they are small, supported on bridge piers, and with the possibility of reducing the foundation points. The base of the bridge pier foundations can be compacted without too great a cost.

There are certain American and Russian publications, which advocate the collapsing of the ground using water before the actual canal construction operations commence, by using pools all along the length of its route. This procedure was employed in the Valle de San Joaquín in California, and in this particular case the ground settled almost 2 m. Some 40 km of this canal were in silty soil.

Problems of collapsing with respect to the presence of silt can be so enormous that in the cases of Terreu Canal, Section III of the Cinca Canal, IV de Monegros and the Pertusa irrigation ditch, all of which can be found in the Ebro basin, the silt was excavated whenever found and replaced with compacted, more or less granular material.

When the canal is constructed, it is necessary to employ all possible means to leave it in such a manner that the water cannot reach the silt, which will require a great deal of attention to be paid to the lining impermeability, its joints and their corresponding filling material.

Since silt does not possess any cohesion once it loses that which it apparently has, and almost no friction either, any slopes constructed in it will not be stable, except where these are very gentle, so that it is not possible to construct a canal with a regular excavation and a surface lining. The strong wall section described in Section 2.6 may be employed, or it is also possible to make use of prefabricated reinforced concrete units, such as those employed in the canals belonging to the Confederación Hidrográfica del Ebro, in the Ebro basin, which can be seen in Figures 12.4 and 12.5.

There is a publication containing a collection of sections, already dimensioned and calculated, corresponding to that of Figure 12.4 (65), which will avoid the need to calculate again.

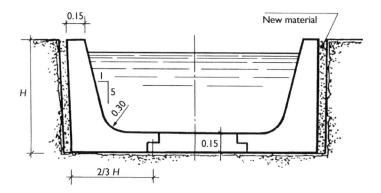

Figure 12.4 Prefabricated pieces (Ebro Canals, Spain).

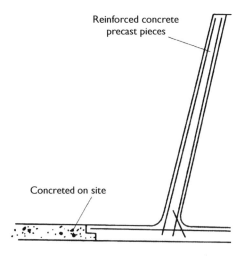

Figure 12.5 Reinforced concrete prefabricated pieces (Ebro Canals, Spain).

12.5 Lining water-filled canals

This is not a rare problem in the world. It occurs when it becomes necessary to repair a canal lining that cannot be taken out of service, generally because it is being used to supply drinking water to a population centre, or where this is one of its multi-purpose uses.

The first solution that might help would be to divide the canal into two parallel watercourses by means of a longitudinal wall. Water is then able to circulate along one watercourse while the other is being repaired. There is an obvious inherent problem with this method because it means that the circulating water flow rate will be a lot less than before. This requires that the flow rate during construction will have to be significantly reduced with respect to the rated value.

Such a repair problem took place in Spain when the Imperial Canal of Aragon, which had been constructed at the end of the eighteenth and beginning of the nineteenth centuries, in permeable ground with a lining of clay soil, had to be repaired without shutting off its service. The numerous vegetation clearing operations performed during the more than two centuries of operation reduced the thickness of the soil lining and the resulting heavy infiltrations made a new lining necessary.

However, the conditioning factors involved in its operation during the 1960s meant that a normal job of work was out of the question. Although the canal was originally intended (13) for the navigation of barges, right from the beginning it was also used for irrigation and for supplying the city of Zaragoza with water. The later use prevented the possibility of closing the canal for long periods and so other methods of lining the canal had to be found without emptying it.

On a test stretch, a system was employed consisting of two lengths of plastic fibre fabric, stitched together as in a mattress. This was laid on the canal bottom and the space between the two lengths was injected with cement and sand mortar. Figure 12.6 shows the resulting lining when the canal was emptied in order to examine the result.

There were several problems to be resolved. The injection had to be carried out at various depths, beginning at the bottom and then ascending up to the banks, and it was also necessary to wait for the lower part to begin setting before injecting the upper part. If it had not been in this way, the very same hydrostatic pressure of the paste, which increased with depth, would have resulted in a defective lining.

The canal's interior profile was carried out using backhoes from both side banks. Since it was not possible to see the bottom, a screen had to be coupled showing the actual excavation profile in comparison with the one designed, as the work progressed.

This system was only employed in a partial lengh of the canal. Almost the same system was employed in the Mogham Canal in Iran (57).

Figure 12.6 Lining constructed under water in Imperial Canal of Aragon (Spain).

In other locations, prefabricated pieces were used and located along the sides in the water to form the canal slopes. The prefabricated pieces can be seen in Figure 12.5.

The Bureau of Reclamation recently constructed a canal lining under water using a machine very similar to the ones described in Section 5.4, but designed to work under water.

One problem in this case is the tendency of the fresh concrete, when it is extended under water, to slide down the slope. In order to reduce this effect, a geotextile was placed between the ground and the concrete.

Chapter 13

Navigation canals

13.1 Introduction

We felt it essential to describe these canals in a separate chapter because they possess certain characteristics that set them apart from other types. While all other canals are dimensioned to carry a given rate of water flow that is necessary for the irrigation of a specific area, to supply drinking water to a population centre, to supply water to a hydroelectric plant for electrical power generation or for the transport of drainage water to a given point, etc., navigation canals require dimensions that are sufficiently large, together with a slow enough water speed to permit navigation along them. It is, of course, required that the canal connects the points that are to be linked for the traffic of goods.

Not all countries or even regions have made equal use of navigation canals. Whereas in Central Europe, including France and England, artificial canal navigation has been very frequent, and also along rivers, in general, optimum conditions for such use have not been available in Spain. This is in spite of the interest shown by its monarchs and governors throughout history. In fact, only two navigation canals have really been constructed, both of which are no longer employed for this purpose: The Castilla Canal and the Imperial Canal of Aragon (Figure 13.1). Both were put into service during the second half of the eighteenth century. On the other hand, in other countries, goods transport by rivers and artificial canals has been extremely useful because of the lower costs involved, together with increased capacity, which was previously much higher than any other method.

The power requirements for goods transportation along navigation canals is approximately one-sixth of that for roadway movement.

In some situations, navigation canals were converted for irrigation purposes, as in the cases of the two mentioned Spanish canals, but, throughout the world, there are still many in operation and interest continues with the construction of new ones under favourable circumstances. These may be solely for transportation use or combined with other uses, producing

Figure 13.1 Imperial Canal of Aragon.

multi-purpose canals, which is why we felt that we could not possibly omit them from a book dealing with canals. In this respect, mention must be made of the Danube-Main Canal, which was completed in 1992, 171 km long, with sixteen locks, each 200 m long, which links the Danube and the Rhine rivers for navigation, so that shipping from the Black Sea can reach the North Sea.

13.2 Route

Apart from general conditions that are common to all canals, such as economic earth movements and avoidance of obstacles, etc., in navigation canals great emphasis must be placed on them being as straight as possible because of the difficulty for the vessels to negotiate tight curves and also to provide the best possible levels of visibility.

For barge traffic up to 400-ton capacity, it is considered necessary that any curves should have a minimum radius of 600 m. For 700-ton barges, this value increases to 800 m, and with 1000 m required for larger craft. These values were obtained from the lengths of barges traditionally employed in Europe (Schleicher).

On curves, the transverse section must be provided with extra width, as occurs with roadways. Its value must not be less than the camber of an arc of circle having the stated radius, with a chord of at least five times the length of the barge.

13.3   Cross section

A navigation canal's cross section must comply with the condition of being sufficiently large for the transit of the expected ship or barge size. This requires, on the one hand, that two vessels are able to freely cross each other, one upstream and the other down, or for one to overtake the other, this means the cross section must be at least twice the width of a barge plus 9 m, allowing 3 m separations between the two barges and between these and the banks.

However, complying with this condition is not enough. It also has to be taken into account that the vessels must be able to navigate against the current, which means a limited water speed that must, of course, be much less than that of the vessel. In a journey along the canal, the barge will occupy a significant part of the transverse section, except in large rivers or canals, and this will push the water upwards increasing its apparent speed, which is transferred along the entire length of the vessel through the narrow space between the vessel and the canal slopes. This requires that the total transverse sectional area to be at least five times the transverse area of the submerged part of the largest vessel. The canal's transverse section has to be large enough to comply with these two described conditions.

In order to prevent the vessel keel from touching the bottom, there must be a minimum free depth of 1 m below under the most unfavourable of conditions.

On many occasions the canal has a trough-shaped transverse section, in other words it consists of a horizontal bottom and slopes in polygon form, with the upper one being steeper than the lower ones. The shape will permit the vessels to come closer to the sides of the canal and also to closely achieve the optimum line of stability for the ground. The lower slopes are usually 1/5 or 1/4 and the upper ones 1/3 or 1/2.5.

The width of the horizontal bottom must be double that of the barge plus 3 m of separation between two barges. There is normally no requirement for any separation between the barge and the side of the canal because the adopted broken slope shape usually provides sufficient space.

One peculiarity of navigation canals is that they require protection on the upper slopes of the transverse section in order to withstand the erosion caused by the waves that are produced by the movement of the barge through the water. This is achieved using a protective layer of rockfill located between levels of +1 m above the maximum level and −1 m. This is another point that differentiates navigation canals from other types. In most, the areas of maximum erosion danger occurs at a point on the slope near the bottom (see Chapter 11). In navigation canals, in addition to this point, the area close to the maximum levels is also a source of danger.

The layer of rockfill usually has a thickness of around 30 cm, with crushed rock of 11–20 cm on the external part and between 5 and 7 cm for the inner part next to the ground. At times, the rockfill is located along the entire length of the slope in order to provide protection against all types of erosion, not just that caused by the barges. In fact, this is the same solution as that proposed for the protection of canals lined with flexible plastic sheet (Section 8.1), but with a greater thickness of aggregates.

It is also necessary in navigation canals to attempt to prevent bottom erosion, but above all in vessels driven by propellers, which can form vortexes. This means that in the case of having not very resistant ground, the bottom should be lined with the same type of rockfill as previously described.

Summing up, from a stability point of view, a navigation canal is unlined or has an improved soil or rockfill lining with all the associated conditioning factors, which is also exposed to wave action and, at certain times, to increased water speeds due to vessel movement.

Combating water losses due to filtration continues to be an extremely important matter. Many navigation canals require a regulating reservoir upstream to provide sufficient circulation flow rates, one very fine example of this being the Panama Canal, which employs water from the Gatún Reservoir. For which reason, all attempts should be made not to lose this precious liquid, but even more so in canals connected to large rivers that are also probably navigable, in which water regulation does not involve any problems of prior investment, however, if the canal is suffering from filtrations, there may not be sufficient water flow downstream in distant areas that are well away from the river. The acceptable loss values are the same as those for all canals (see Chapter 2).

The fight against filtration may make it necessary to employ an improved soil lining with high percentage of clay, underneath the protective rockfill lining. Never should it be forgotten that, in most cases, a navigation canal is an unlined canal, and therefore the same solutions are frequently applicable (see Chapter 11).

13.4 Locks

These are a very typical element along navigation canals. They can be thought of as steps distributed along the canal so that its slope is not as steep as that of the land through which it runs because of the slow water speed they require. The canal therefore consists of a series of canal sections each having a gentle slope, separated by the locks.

Locks are canal sections having a length sufficient to hold one or two of the expected largest size of vessel, one behind the other, and also with sufficient width for one or two side-by-side vessels.

Locks which produce a drop or level change in the canals are fitted with connection gates at both the upstream and the downstream ends. These gates are large enough to allow free passage of the vessels when they are opened. In addition to these main gates, there are also other smaller ones that are used to fill the lock from the upstream section or to empty it into the downstream section.

The vessels are able to navigate normally along the stretches between locks. When they arrive at one of these locks, its level is made to match that of the canal section in which the vessel is currently in. The main gates are opened to allow the vessel pass inside. The level inside the lock is changed to correspond to that stretch of the canal to which the vessel intends to go, by either raising or lowering it. When this is achieved, the vessel is able to continue normal navigation along this following stretch.

In a fashion, locks act as lifts that allow the vessels to ascend or descend from one section to another on a different level. This involves the transfer of a volume of water from the upper section to the lower one, first to fill the lock from the upper section and then to empty it into the lower one. This is precisely the flow rate control system required for a navigation canal.

Locks normally have a rectangular bottom section, with sides in the form of vertical walls to facilitate the entrance of vessels having different draughts. These walls are traditionally constructed using mass concrete, but with greater-than-normal thicknesses because there may be significant uplift pressure.

Lock bottoms may also be subjected to very significant uplift pressures due to the difference between its level and that of the upstream section, which may suffer filtration that soaks the ground and pushes the bottom upwards when the lock is emptied.

For this reason, the bottoms are sometimes designed in the form of an inverted vault in order to resist these forces. Otherwise it may be necessary to consider the possibility of employing drainage underneath, using the techniques described in Chapter 9.

A lock is an expensive construction and so its dimensions should be kept as small as practical. Its width is often less than that of the canal, with a reduction of the free space between the barges and between the barges and the side walls from the previously stated 3 m for the canal to only 1 m. The lock length should be the maximum expected barge length plus another 5 m. The 1 m free depth under the keel is maintained at 1 m; however, these should be treated as merely guideline values. Lock freeboard should be a minimum of 0.5 m above the maximum expected level.

Each lock should include side gates for canal regulation because there are usually many occasions on which, due to a lack of vessels or excess circulating water, this should be released through the regulation gates. Adequate implementation is shown in Figure 13.2, in which can be seen the lock and the regulation and the spillway system in each arm.

Lock location should be very carefully selected. It is evident that the most suitable locations are those with local differences of level in the terrain covered by the canal. These differences in level will be fixed according to topographic reasons. In general, however, it is necessary to construct

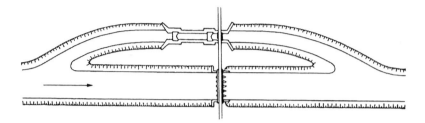

Figure 13.2 Lock implementation.

additional locks in order to reduce the effects of the canal slope. The further apart the locks, the greater the difference in levels that has to be overcome. This obviously makes it essential to perform a comparative economic study. It might be beneficial to select equal drops, not only to standardize elements, for example the gates, but also for conditions deriving from the resulting flow rates, which will be examined in the following section.

In principle, the gates the vessels pass through do not have any limitations, other than allowing them to pass, which imposes significant conditioning factors if theses vessels are required to pass underneath them when they are open.

Rolling gates have been widely employed. These consist of two-leaved gates that rotate on vertical shafts and when closed, one rests on the other. Its resistance calculation is simple, taking into account that the reactions of the end of one against the end of the other, for reasons of symmetry, have to be equal and both in a perpendicular line to the lock axis. Since the hydraulic forces acting on the gate are known, it is quite easy to obtain the direction and magnitude of the reaction on the rotation axes through a breakdown of forces.

13.5 Flow rates, slopes and speeds

One basic piece of information required for the dimensioning of a navigation canal is the type and number of vessels it will have to carry daily.

For guideline purposes, European barges are some 55 m long, 7.4 m wide, together with a draught of 1.4 m and transport capacity of 400 tons.

At 600 tons, these values are increased to 65 m long, 8 m wide and 1.75 m draught.

For 1000 tons, these then become 80 m long, 9.2 m wide and 2 m draught.

The vessels expected on the previously cited Spanish canals are much smaller.

Once the maximum expected length is established, a prior dimensioning may be made for the cross section in accordance with the given rules.

The locks are also to be dimensioned with respect to the largest expected vessels. This is a very important piece of information because each time a barge enters a lock, it is necessary to empty it and fill it again, or viceversa. The volume of water involved in this operation is the surface area of the lock multiplied by the difference in water levels between the upper and the lower sections of the canal. The released flow obviously depends on many factors, such as the water energy to be dissipated and the produced waves, etc. But in any case, no matter how small the lock is, never less than ten minutes should be allocated to the emptying operation and the same amount of time for filling, although it is not possible to provide any fixed rules. For example, the Panama Canal, which is much larger than the navigation canals referred to in this book, has an approximate length of 80 km and includes eight locks which are each 300 m long. The average time required for a vessel to cross the Canal is eight hours, which only leaves a very small time for lock operations.

The circulating flow through the canal is therefore defined by the volume of a lock divided by the total fill and empty times plus the losses due to filtration and evaporation, etc.

When there are several locks along the length of a canal, the flow released by one upstream will arrive later and successively at each of the locks downstream. If all these locks are exactly the same, the flows released by each one would also be the same and the canal would operate with a constant flow rate along its full length. However, in a normal situation, there are locks with greater volume, for example, because they have a greater difference in level to overcome. The corresponding flow will be greater than others so that it has to be released through the regulation systems. It can be understood that in the interests of economy, the ideal solution would be for all the locks to have the same volume, together with the same systems for closure.

It is then necessary to select the acceptable velocity for the canal and the corresponding longitudinal gradient. Obviously no erosion should be produced, but the canal must also permit safe navigation, which means the speeds have to be low. For reference purposes, vessels crossing the Suez Canal are limited to maximum speed of seven knots.

With a maximum water speed of 1 m/s, which is the equivalent of two knots, a vessel travelling at five knots, which is recommended for artificial navigation canals, will advance at three knots, in other words, at 5.4 km/h or 1.5 m/s, which is possible with the usual haulage means along a tow path (Section 13.6). Nevertheless, we are of the opinion that a maximum water speed of 0.50 m/s would be a much better situation. The corresponding slopes are calculated using Manning's Formula with a coefficient of 0.0023, taking into account the rockfill protection. If the cross section turns out to be insufficient, then it will have to be corrected in order to obtain a flow rate at the desired speed.

It is also necessary to plan for moorings so that, during times of excessive traffic, vessels waiting for their turn to enter the locks can be tied up. These mooring areas are usually wider than the actual locks themselves.

13.6 Tow paths

The vessels making use of navigation canals can be self-propelled, but on other occasions, they are towed either by a tug or, especially on small canals, towed from the canal side. The towing force in the latter case was traditionally provided by mules, but this was subsequently replaced by mechanical tractors. In both cases, ropes were used to pull the barge.

A track is required on either side of the canal for the mules or tractors, and this is known as a "tow path". A minimum width of 2 m is required for animals and 3.5 m for tractors.

The standard height of the tow path over the maximum water level is usually 3.5 m. If there are any bridges over the canal, these must obviously have sufficient clearance to allow the vessels to pass underneath and also sufficient span to include the tow path. The normal free height between the bridge and the level ground underneath is normally 3.5 m or more.

The tow path is always a necessary feature at locks, even when the vessels are self-propelled because their manoeuvrability inside the lock is difficult as its short length does not permit the use of their engines. Inside the locks, the vessels are moved and steered from the tow path. One example of this can be found in the Panama Canal, where the vessels are handled in the locks by the so-called "mulillas" (meaning small mules), a traditional name given to small, but powerful tractors running along racks.

13.7 Multi-purpose canals

Hydraulic constructions, among them canals, may have one or several uses. These uses may be for consumptive purposes or not. Within the former type, the use is mainly for irrigation, which uses up the water through evapotranspiration, and water supplies for human consumption, which, although a significant part is returned, is contaminated. The two main non-consumptive uses are hydroelectric power generation and navigation, neither of which actually consumes the water, but they do return it at a lower height.

Because of their very nature, the association of two consumptive uses reduces the water availability for both. On the other hand, consumptive uses may be combined with others that are not. The most frequent case is that of combining hydroelectric power production, which obviously includes a drop or level difference, with the supply of drinking or irrigation water, for example, in the case of regulating reservoirs.

Figure 13.3 Cacique Guaymallén Canal (Mendoza, Argentina).

In the specific case of canals, it is at least theoretically possible to associate hydroelectric drops with navigation canals, by locating the drops and hydroelectric plants next to the locks, thus taking advantage of the existing differences in levels and flow rates in excess of those necessary for navigation. In fact, there are mills and hydroelectric plants located at the locks along the Castilla Canal.

When a new canal is being studied and designed, it would be a good idea to verify whether it would be worthwhile increasing the transportable flow rate if any is available, or to vary certain specifications in order to employ it for more than one use.

This, for example, would be the case of a navigation canal ending in the city to which it is used for goods transportation. Perhaps the water could be employed in the irrigation of nearby areas, where the agricultural production could be very significant to the population of the city. A great number of possible cases may be imagined. Evidently, in each case, the canal will have to be dimensioned to include all the specifications required for each individual purpose, such as flow rate for irrigation purposes, and sufficient canal section for navigation, etc.

One infrequent case of a multi-purpose canal is that of the Cacique Guaymallén Canal in Mendoza, Argentina, shown in Figure 13.3. It is used to evacuate flood waters and that from the city drains and also to transport irrigation water. Since flood water evacuation and irrigation operations are not simultaneous uses, their combined use is quite compatible.

Chapter 14

Watercourse crossing works

14.1 Elevated flumes

These are structures that are similar to bridges and transport the canal water over rivers, other watercourses and obstacles.

14.1.1 Purpose

Canals normally have only very small gradients and this means that their most economical route is usually almost coincident with a contour line, in relation to which it descends a few centimetres every kilometre at the same time as slightly separating from it.

However, there are times when the canal route meets a watercourse that presents the design engineer with two possibilities. One is to follow the same contour line, bordering the watercourse. The other is to cross it by means of a structure, which is really a bridge, known as an elevated flume, supporting the canal. This second option opens up a whole range of variants since the crossing point can be located over the watercourse either further upstream or further downstream.

It is, in any case, essential for this new construction to allow the passage of rainwater that comes along the watercourse and has to cross the canal, either above or below it.

If the flume solution is adopted, very careful attention should be paid to the fact that most roadway bridges that have collapsed throughout history have done so due to insufficient capacity to drain away the river or watercourse waters. If the available section under the flume is insufficient for the natural passage of the waters, then the flume will be loaded with a water level upstream from the flume that is higher than the downstream level. Pressurized water will then circulate under the flume at high speed, which will lead to undermining of the pile foundation with subsequent general collapse.

If it is decided to adopt the solution of crossing the watercourse with a flume, then its location, either upstream or downstream, will have to be

Figure 14.1 Structure for passing water over the Carrizal Canal.

chosen so that it has sufficient capacity to allow safe passage of the water carried by the watercourse.

Other devices are required if it is decided to completely route the canal alongside the watercourse. At times, an overpass is constructed over the canal, which will transport the rainwater arriving via the watercourse to the lower section. This can be seen in Figure 14.1, which shows the existing construction for the Carrizal Canal in Mendoza, Argentina.

On other occasions, preference is given to one or more pipes passing underneath the canal, provided that the only moderate flow rates are to be carried. A syphon inlet is employed, but it is very important for the outlet to be as free as possible, without the water having to overspill via a syphon outlet because the watercourse waters will be carrying stones and sediment that could block the pipes.

At times, when the watercourse flow rates are very small, water is allowed to enter the actual canal itself, using the canal freeboard for this purpose.

14.1.2   Flow rate evaluation in watercourses

In most cases there is no streamflow information available, which means rainfall data has to be employed, together with a hydrological study of the flow rates in the contributing basin associated with the design.

It is usually enough to carry out a survey similar to those performed in motorway projects. We believe that the time of concentration or unit hydrograph methods are both acceptable.

It is first necessary to select the time of concentration, which is the time in which it is assumed that the drop of water which is most hydraulically distant requires to reach the area of the flume. The value given by the following formula may be employed:

$$T_c = 0.3(L/J^{1/4})^{0.78}$$

from the US Army Corps of Engineers, in which:

L is the distance between the water drop and the elevated flume along the watercourse in kilometres
T_c is the time of concentration in hours
J is the average slope of the main course.

The formula that gives the maximum water flow is as follows:

$$Q = \frac{C \cdot I \cdot A}{3.6} \cdot K$$

where

Q is the flow rate, in cubic metres per second
C is the runoff coefficient
I is the average hourly rainfall intensity during time T_c, in millimetres per hour
A is the basin surface area, in square kilometres
K is a correction factor for irregular rainfall distribution which, according to Témez (91), may be taken as being 1.2.

The adopted value for C is always open to discussion because, within the same area, it can vary from zero, for rainfall that is less than the minimum required for runoff, up to one during a downpour. It will also depend on whether the ground was already wet or not, etc.

Always in accordance with José Ramón Témez, the value provided by the following formula may be used for C:

$$\frac{(P_d - P) \cdot (P_d + 23 \cdot P_0)}{(P_d + 11 \cdot P_0)^2}$$

where P_d is the daily rainfall of the situation under analysis and P_0 is the runoff threshold, in other words, the minimum rainfall value required for runoff to occur. If P_d/P_0 is less than one, the runoff coefficient is evidently zero.

The value of P_d should be adopted with respect to the information provided by the meteorological observatory, and the value of P_0 can be taken from the following tables:

Runoff threshold P_0

Soil use	%Slope	Type	Soil group			
			A	B	C	D
Fallow	≥3	R	15	8	6	4
	≥3	N	17	11	8	6
	<3	R/N	20	14	11	8
Crops-row	≥3	R	23	13	8	6
	≥3	N	25	16	11	8
	<3	R/N	28	19	14	11
Winter cereal	≥3	R	29	17	10	8
	≥3	N	32	19	12	10
	<3	R/N	34	21	14	12
Rotation of poor crops	≥3	R	26	15	9	6
	≥3	N	28	17	11	8
	<3	R/N	30	19	13	10
Rotation of dense crops	≥3	R	37	20	12	9
	≥3	N	42	23	14	11
	<3	R/N	47	25	16	13
Poor meadow	≥3		24	14	8	6
Average meadow	≥3		53	23	14	9
Good meadow	≥3		*	33	18	13
Very good meadow	≥3		*	41	22	15
Poor meadow	<3		58	25	12	7
Average meadow	<3		*	35	17	10
Good meadow	<3		*	*	22	14
Very good meadow	<3		*	*	25	16
Poor forest plantation	≥3		62	26	15	10
Average forest plantation	≥3		*	34	19	14
Good forest plantation	≥3		*	42	22	15
Poor forest plantation	<3		*	34	19	14

(Continued)

Runoff threshold P_0

Soil use	%Slope	Type	A	B	C	D
				Soil group		
Average forest plantation	<3		*	42	22	15
Good forest plantation	<3		*	50	25	16
Very open forest mass			40	17	8	5
Open forest mass			60	24	14	10
Average forest mass			*	34	22	16
Thick forest mass			*	47	31	23
Very thick forest mass			*	65	43	33

In this table R indicates that the crop is grown in direction of the maximum slope and N means it follows the contour lines. The symbol "*" indicates that this basin must be considered as non-existing with respect to the calculation of spate flow rates.

This table requires the insertion of values for *A*, *B*, *C* and *D*, which are selected in accordance with soil classification provided below:

Soil classification

Group	Infiltration	Depth	Texture	Drainage
A	Fast	High	Sandy Silt-sand	Perfect
B	Moderate	Medium to high	Loam Sandy loam Sandy-clay Loam Silty loam	Good to moderate
C	Slow	Medium to small	Clay loam Silty-clay Loam Sandy Clay	Imperfect
D	Very slow	Small (clay subsoil) or horizon	Clay	Poor or very poor

The previous tables provide the P_0 coefficient for vegetation-covered soils. When dealing with rocky or paved ground, the following table must be employed:

Ground type	Slope (%)	P_0(mm)
Permeable rock	≥ 3	3
	< 3	5
Impermeable rock	≥ 3	2
	< 3	4
Granular unpaved roadbed		2
Paved		1.5
Bitumen or concrete paved		1

In order to apply the time of concentration formula which does not give the maximum flow rate to be considered, the value of I must be defined.

Here I is the average rainfall in time T_c, which is chosen within a downpour in order to obtain the largest value of I, it can be established from a meteorological station data, assuming there is one close enough, and can be calculated using formulas of the type: $I = a \cdot t^b$, in which a and b are parameters that should be taken from data for other downpours and which vary from one region to another. The Temez formulas (91), which are not so well known, may also be employed.

The formula that provides the maximum rainfall in function of the time of concentration is only valid for surfaces that are less than a rather small value of only a few square kilometres, because it ignores the water storage effect in the ground and assumes that the runoff coefficient is constant during the time period T_c under consideration. For larger basins or those cases requiring greater precision, it will be necessary to employ more exact methods, for example the unit hydrograph, and also to consult books on hydrology, some of which are especially recommended (64), (97).

Everything stated to this point refers to rainfall and its subsequent flow rates which occur on a frequent basis.

However, experience shows that there are dry years and wet years, with some being especially rainy. This same experience also shows that the greater the number of years that are taken into account, the higher are the maximum rainfall values.

In general, it can be said that, once a long period of years has been selected, there are only a certain number of times that a specific rainfall occurs, which leads to the association of the maximum rainfall with the probability of its occurrence, with the rainfall being greater with less probability.

The probability of a specific rainfall is defined as the number of times it occurs in a large number of years divided by this same number of years.

The inverse of this quantity is called the "recurrence period", which is similarly related to the rainfall intensity to such a point that it is quite normal to define it in terms of the recurrence period.

The choice of rainfall recurrence period to be taken into account in the dimensioning of a flume project works is transcendental, because if it is very low the rainfall will occur with disturbing frequency and flood damage, and if it is very high the construction works may be prohibitively expensive.

It is therefore necessary to know how to relate annual rainfall data, which are already known, with those that are probable with respect to variable recurrence periods of ten, one hundred and five hundred years, etc., which are those affecting our civil works for sewers, flumes or dams and thus be able to produce a suitable design. This requires the use of statistical methods.

There are several laws that are very frequently used quite successfully, such as those of Pearson and Gibrat, etc., although one of the best-known ones is that of Gumbel.

If x is the magnitude of a given rainfall, the probability of it actually occurring is given by Gumbel's Law as follows:

$$f(x) = e^{-e^{-a(x-x_0)}}$$

where a and x_0 are two parameters.

The probability of this phenomenon occurring will be $1 - f(x)$, and the recurrence period is $T = 1/(1 - f(x))$.

Once these two parameters, a and x_0, are established, by means of the known data series that relate the various rainfalls with their associated probabilities of occurrence, the function $f(x)$ is defined and hence T also, which will enable us to relate each maximum rainfall value with its recurrence period T.

Since the function $f(x)$ is complicated, another method which is easy to follow is the graphic procedure described below.

If a change to an auxiliary variable $y = a(x - x_0)$ is made, on Cartesian axes x and y this relationship will be a straight line. In addition, the function $f(x)$ will be transformed into another $f(y)$, as shown below:

$$f(y) = e^{-e^{-y}}$$

As can be seen, x does not take part in this and so we are able to define it, once and for all, by placing the value of $f(y)$, which is the probability of a determined rainfall occurring, on the Cartesian axis alongside each value of y.

However, the period of recurrence T is also expressed, without the participation of x, by the expression $T = 1/(1 - f(y))$, so that it also can be

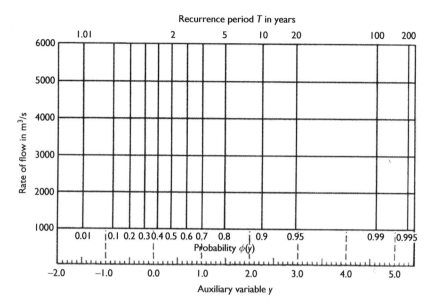

Figure 14.2 Graph for working with the Gumbel law.

indicated, once and for all, by putting the corresponding value of T on the Cartesian axis, alongside each value of y.

The result is a graph, such as that shown in Figure 14.2, in which x is given on one axis and the other represents y, $f(y)$ and T simultaneously.

The successive pairs of specific known rainfall values (x), together with their recurrence periods (T), define a series of points drawn on the graph. A straight line is adjusted to fit these, which represents the relationship between x and y. Once this has been defined, the graph can then be used to establish the values of maximum rainfall x corresponding to each period of return T.

The straight line can be adjusted to fit between the average of the points, but it would be even better to use the mathematical method of minimum squares, in which case there will be no need to graphically adjust the straight line.

The Gumbel method is widely employed throughout hydraulic engineering, above all for determining spates with long recurrence periods. Simpler formulas may also be employed in the design of flumes, such as that of Fuller.

Fuller developed his formula, not for rainfall study, but instead for spate flow rates. The formula is given below:

$$q(t) = q(1) \cdot (1 + 0.8 \log_{10} t)$$

where $q(t)$ and $q(1)$ are the recurrence period flow rates for t years and one year respectively.

The spate can be calculated for a short recurrence period for which we possess data, and this formula can then be used to extrapolate for longer periods. Its simplicity enables us to calculate that the maximum flow rates for various recurrence periods are as follows:

1 year	q
2 years	$1.24q$
5 years	$1.56q$
10 years	$1.80q$
25 years	$2.12q$
100 years	$2.60q$

These values will be of great use when the time comes to decide on the recurrence period to be used in the calculation of a flume.

In hydraulic works design it is necessary to select the return period for the design spate. When a large factor of safety is required, a long period should be chosen, to which a greater spate is associated. For example, a minimum of five hundred, one thousand or even ten thousand years is chosen in dam spillway design. This is because the loss in human lives and material damages would be very high should a dam break.

On the other hand, a return period of ten years, for example, may be selected for the design of a city's sewer network, because in a situation of heavier rainfalls that exceed the sewer capacity would flood the streets and cause a certain disturbance to the populations, but would not cause such a large amount of damage.

In the case of a flume that collapses, irrigation would be impossible during a certain length of time, which would obviously lead to significant economic losses, but not as much as in the case of a broken dam. There are no clear regulations as to which return period should be selected for flume design, but in view of the stated considerations we would recommend a period of one hundred years.

14.1.3 Types of flume structures

During the last century, flumes were constructed in the same way as roadway bridges. They generally consisted of vaults on top of which the corresponding filling was applied followed by the actual canal itself instead of the roadway or railway (Figure 14.3).

Subsequently, with the appearance of reinforced and pre-stressed concrete came the possibility for the canal itself, in addition to transporting water, to be a resistant element working in the same way as a beam, for example.

The first solution is evidently formed by canal sections consisting of the bottom and the lateral walls, which act as an isostatic beam supported

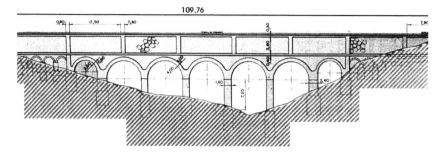

Figure 14.3 Flume on top of vaults.

on piles. This is not, in fact, a good solution because, with this layout of isostatic beam, the concrete section which has to withstand the compression forces is the upper part of the lateral walls, thus wasting the opportunity of using the bottom, with its larger section, for this purpose.

A much better solution is the "T" beam, in which each pile is monolithic with two cantilevers that are half of each adjacent span. This implies that there is a joint at the centre of each span Figure (14.4). This is the solution adopted in a flume for the Trasvase Tajo-Segura Canal. It has the advantage of the cantilever compression forces being in the bottom area, which is a zone with a strong concrete section that is very suitable for withstanding such forces. On the other hand, the upper parts of the lateral walls undergo

Figure 14.4 "T" type flume in the Tajo-Segura water transfer.

traction forces and this is where the reinforcement steel has to be employed. For this reason, two cantilever catwalks are constructed, which, in addition to permitting access for maintenance work, also make suitable housing for the reinforcement steel.

Another similar solution is the so-called "π beam", which consists of a length of canal that is twice that of a span, and acts as a supported beam, but with a half-span-length cantilever on each side. The bending moments at the cantilever origins are equal and opposite to the moment at the centre of the beam considered as support and without cantilevers. This means that the final law of moments is that shown in Figure 14.5 and the entire bottom is in compression, which allows maximum advantage to be made of the concrete's compression-resistant quality. Again in this case, two catwalks are constructed for the double purpose of maintenance and traction steel housings.

Reinforced neoprene sheets should be positioned over the beam supports on the two piles, which will combat rotation and translation caused by mechanical and thermal loads. This solution, in the same way as the "T" beam, is isostatic and is very favourable in the case of differential settling of the piles. This solution was adopted in the Orellana Canal, Alcollarín Flume (Figure 14.6).

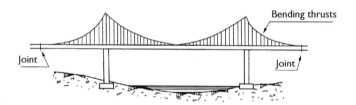

Figure 14.5 Thrusts in a "π beam" flume.

Figure 14.6 Alcollarín Flume (Spain).

On occasions, a box-beam is adopted for the flume section. In addition to forming a strengthening element, the upper part can also be used as a highway or service road, providing it has sufficient dimensions.

The side walls of a flume have to be able to withstand the hydrostatic thrust produced by the contained water. The most economic method of supporting them is to brace one against the other by means of cross reinforced concrete beams (Figure 14.7).

However, this is not always carried out as clearly shown by the famous Tardienta Flume (Figure 14.8).

A rectangular or trapezoid transverse section is usually adopted for the flume, with slopes that are very close to vertical (Figure 14.4). It is usually recommended that the sections be higher than their width in order to obtain a greater strength resistance.

Figure 14.7 Flume with side walls braced by reinforced concrete beams.

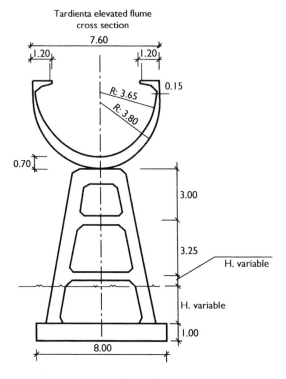

Figure 14.8 Tardienta Flume (Spain).

Up to this point we have dealt with flumes constructed using reinforced concrete. However, pre-stressed concrete also finds use in the construction of flumes, as can be seen in Figure 14.9 which shows the Alcanadre Flume on the Cinca Canal. This flume is also significant because of the construction method consisting of pushing and transferring the various sections that were constructed previously (Figure 14.9).

If the resistant structure is also used for transporting the water, the flume may be cheaper. This advice is not always followed, as for example in the Gargaligas Flume, Spain (Figure 14.10).

14.1.4 Hydraulic study of flumes

Flumes are expensive structures and for this reason attempts are made to reduce costs by increasing the slope to augment speed and reduce the transverse section. In the study of a canal slope close to a flume, a greater difference in level should be included in the plans in order to take the increased gradient into account.

Figure 14.9 Alcandre Flume (Spain) constructed with pre-stressed concrete.

Figure 14.10 Gargaligas Flume (Spain).

The water depth at the flume entrance will undergo a descent. The arrival kinetic energy may be ignored because, apart from being small, it is not recovered in practice. There is always some loss at the entrance due to the change in section, which we can assume to be $0.3V^2/2g$, where V is the velocity in this section of the flume. The water depth descent at the flume entrance is therefore equal to $1.3V_2^2/2g$, with V_2 being the speed in the flume.

With the new water depth height established and a first calculation for the water height carried out, it is now possible to calculate the level of the bottom at the flume entrance.

Depending on the actual situation, the velocity within the flume is either greater or less than the critical value. At the structure's exit, where it joins up with the canal's normal section once again, a water level curve will be formed of type M or S (see Chapter 1), and it is highly recommended that a study be carried out in accordance with the contents of this chapter and sufficient knowledge of hydraulic theory. The kinetic energy that is theoretically recoverable is $V^2/2g$, where V is the water speed in the flume and ignoring the small amount of energy in the downstream section of the canal. Nevertheless, no matter how well the transition is designed and constructed, there is always a loss of about one half. Only $0.5 \times V^2/2g$ is recovered, so that the water depth height on arrival has to be that of the following canal section minus this amount. Since we have calculated the water height, the bottom level at the flume entrance and exit is given by the difference between the water level and the corresponding water depths.

The flume freeboard is dimensioned using the same criteria as normal sections, taking the precaution that the upper edge of the flume is never at a lower level than the following canal section. In this way, overflowing will be prevented when the water depth situation is altered because of a reduction in speed.

After all these calculations have been performed, it is possible that the flume slope and cross section have to be recalculated in order to fine-tune and adjust the obtained numbers.

14.2 Inverted syphons

Inverted syphons are works that are inserted into canals in order to cross underneath a specific obstacle, which is usually a roadway, path, railway or even a watercourse or river.

They basically consist of one or more pipes that are nearly always buried, through which the water passes under pressure, although in many cases this is small, together with two connections, one at the entrance and another at the exit of the canal. These works are also known as syphon inlets and outlets or transitions.

Whenever possible, these pipes should be prefabricated, for improved quality assurance, with watertight seals supplied and guaranteed by the manufacturers. Although fibrocement and PVC can be employed for small diameter pipes, in general concrete is employed. This can be mass concrete for diameters up to 60 cm, because the internal pressure is often low. However, it is recommended that they are made of reinforced concrete, which is in fact necessary for larger diameters in order to guarantee sufficient strength against crushing pressure produced by the soil loads and any fixed or rolling overpressures.

Due to the fact that the pipe is more expensive than the section of canal it replaced, the water speed is often increased in order to reduce the diameter somewhat and hence, its price. This can also make it possible to employ standard commercial sizes, which do not usually exceed certain values.

If the flow rate to be carried is too large for commercial pipes, instead of ordering the production of large diameter piping, it might work out more economical to install two pipes selected from those available. This will mean that the transition or syphon inlets and outlets will be wider, something that will have to be taken into consideration, when carrying out comparative studies.

Above a certain flow rate, it is no longer possible to employ prefabricated pipes, except in the case where the total syphon length is so long that it becomes economically feasible to order the production from a specialized concrete plant. In other cases, there is often no other possibility than to construct the pipes onsite.

Traditionally, circular section pipes are used but there are no hydraulic or constructive reasons preventing the use of other sections. In fact, for inverted canal syphons crossing roadways and paths, where the pressure is usually very low, the recommended section is rectangular, whether one or more pipes are employed (Figure 14.11).

Among its advantages, emphasis should be placed on the fact that, since all its interior and exterior faces are flat, the associated formwork will be both simple and economical and may be prepared by personnel with very little specialized experience.

It should also be pointed out, that the rectangular shape employs the total width for water transport, something that is not achieved with a circular section, which does not use the four vertices. This means that for the same capacity, a square inverted syphon would be narrower than a circular one and hence the syphon inlet and outlet works will also be both narrower and economic. If we now include the fact that the recommended shape for a rectangular inverted syphon is not square, but narrower and higher, the savings in transitions are even more noticeable.

From the design and construction points of view, there is no reason for rectangular syphons not to be used in most situations. Although it is true that a circular shape withstands interior pressure better, it is no less certain

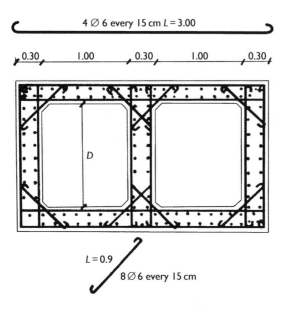

Figure 14.11 Rectangular cross section for syphon pipelines.

that the pressure in most canal syphons is very small (there are exceptions with very significant pressures that have to be treated as high-pressure piping installations), which can be easily withstood by the flat piping walls, provided they are constructed using reinforced concrete. A similar statement can be made with respect to soil loads and overpressures.

In the case of flow rates in excess of a certain value, it is better to employ several pipes in place of just one. This will reduce the interior spans and facilitate strength against soil loads and interior pressure. It is recommended that the interior corners of the rectangle are bevelled, since these are mechanical force concentration zones where the material has a tendency to crack.

The thermal stresses produced in the pipes due to environmental temperature variations and, above all, to the filling or emptying of water must also be taken into account. These stresses require longitudinal reinforcement steel in the reinforced concrete pipes to absorb the resulting traction forces.

When the piping is long, it will be necessary to cut it into shorter lengths with intermediate joints. The actual length will be given by the structural calculation, but we believe that it should not exceed approximately 8 m.

The type of joint employed should be very carefully decided, with one of the most logical types for onsite-constructed rectangular pipes being impermeable adhesive strips stuck on both sides inside the pipe, having previously left a small slot on the concrete so that it does not protrude

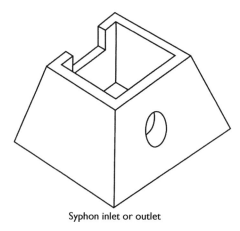

Syphon inlet or outlet

Figure 14.12 Simplest and traditional shape of syphon inlet or outlet.

beyond the facing. It is obviously assumed that the pipe is accessible so that the joints may be inspected and repaired where necessary.

Inverted syphon transitions or inlets and outlets are very important elements. The simplest traditional form, used in small canals or ditches, is shown in Figure 14.12. The facings are flat, both interior and exterior, which means they can be easily constructed using bricks or concrete, without the need for highly specialized personnel.

If the difference in water level between the water entrance and the exit of an operating inverted syphon is measured with a topographic level, it will be seen that there is usually an appreciable difference of several centimetres, which is the head loss produced by the inverted syphon. Irrigation canals are often located on flat areas where there is no possibility of adapting significant head losses, which means careful prior study is required. Figure 14.12 requires the water to follow a sinuous path, which will produce large head loss values. This has resulted in the design of other type of inlets and outlets, such as that shown in Figure 14.13, which provides a less complicated water flow, with consequently less head loss. At the same time, an attempt has been made to reduce its form and construction, with all faces, except the bottom being flat.

The prior calculation that pays great attention to the head loss of a future inverted syphon for a canal is very important. The difference in level at the site must be equal to the expected head loss for the maximum design flow rate. This means that if the transversal sections and water depths are equal both before and after the inverted syphon, then it will be necessary to construct the bottom with a difference in level equal to the calculated

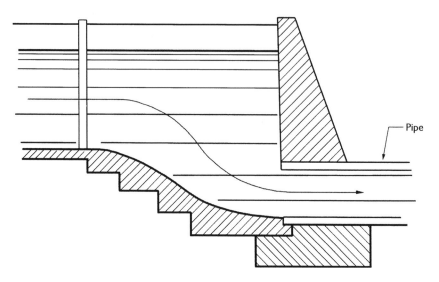

Figure 14.13 Syphon inlet or outlet with less head losses.

head loss. If not, Bernoulli's Theorem will have to be applied to calculate this difference in level.

If *P* is the head loss, *s* the bottom drop between the syphon inlet and outlet, y_1 and y_2 are the water depths and v_1 and v_2 are the speeds at the inlet and outlet respectively, then this theorem will give:

$$s = y_2 - y_1 + V_2^2/2g - V_1^2/2g + P$$

The head loss in an inverted syphon consists of many different summands. The most important ones are as follows:

1   Transition losses between the canal and the syphon inlet and outlet and within these
2   The difference between the creation of speed in the pipe (greater than that of the canal) and that which is recovered at the exit
3   Losses at the pipe inlet and outlet
4   Friction losses inside the pipe.

The energy losses due to the creation and not recovery of speeds within the syphon inlet and outlet can usually be ignored and have not been mentioned for this reason, however under certain circumstances they may have to be taken into account.

Among those described, the most important ones are those indicated in points 2, 3 and 4 above.

The summand described in point 2 is generally taken as being completely lost because it would require a perfect transition from the hydraulic point of view between the pipe exit and the final syphon outlet, something that does not actually occur in practice. If V_t is the speed inside the pipe, this will be $V_t^2/2g$.

The pipe entrance and exit losses greatly depend on whether it has trumpeted ends or not. Between one possibility and the other, the losses may vary between $0.08V_t^2/2g$ and $0.5V_t^2/2g$.

Since the pipe is almost never trumpeted, the maximum value is usually taken into consideration which, together with the creation of speed inside the pipe, comes to $1.5V_t^2/2g$.

However, under certain special circumstances, this value can be somewhat reduced, for example, if the pipe end is rectangular and the syphon inlet and outlet are the same width as the pipe so that there is no lateral water width contraction.

The losses described in point 1 above may be substantially reduced by careful transition design and because of this, they are not generally taken into consideration.

Lastly, there are friction losses described in point 4 above, which are proportional to the pipe length. They can be calculated by means of any of the Manning, Kutter or similar formulas, by introducing the speed value, which is known because we have established the maximum flow rate and the area of the cross section, together with the pipe rugosity. The formula gives the loss which is called I. If the pipe length is L, the loss will be $I \cdot L$, which finally becomes the simplified, but quite approximate, formula for the total head loss in the syphon:

$$1.5.V_t^2/2g + I \cdot L$$

In irrigation canals employing syphons to cross roadways, etc., the pipe length is very short, and so the second summand disappears.

The same does not occur when the syphon pipe crosses a watercourse or similar obstacle, in which case this summand may become extremely significant.

In the same way, this last case may involve elbows or curves in the pipe, which will produce additional losses that will have to be calculated by means of the usual formulas which are included in the forms for such cases.

For example, the Weisbach Formula (46) could be employed

$$H = \xi \cdot \left(\frac{V^2}{2g}\right) \cdot \left(\frac{b}{90°}\right)$$

where

H is the head loss
V is the water speed
g is the acceleration due to gravity
b is the angle at the centre of the curve
ξ is given by the expression:
$\xi = 0.131 + 1.848 \cdot (r/\rho)^{7/2}$, where r is the curve radius and ρ is the pipe radius (assuming it is circular).

If the pipe is rectangular and s is the inside pipe width, we obtain:

$$\xi = 0.124 + 0.274 \cdot (s/\rho)^{7/2}$$

14.3   Tunnels

14.3.1   Why they are needed? Shapes of Cross sections

Among the obstacles that may be found when establishing the route for a new canal are mountains or simply an area of higher levels which, to cross in the open air, would require extremely deep cutting.

In such situations, it may be recommendable to overcome the obstacle via a tunnel, lined or unlined, pressurized or not.

Tunnels are also usually necessary for the transfer of water between different basins.

In general, except in the case of pressurized tunnels, stability design for hydraulic tunnels is the same as for any other type of tunnel, hence the same methods are applicable.

On the other hand, the criteria employed for the shape of the cross section may be very different.

When canal design engineers find that the topography imposes that part of the canal has to be constructed inside a tunnel, they continue with the same idea in mind and treat it more or less as a regular canal covered with a vault. Of course, they will not choose a trapezoid transverse section with the same transversal slopes as a canal since this would lead to a vault with a large, very expensive span and instead adopt a rectangular, or almost rectangular section, with 1/10 slopes, with the vault being supported on the lateral walls, which then become side walls.

This transverse section has been used to construct a very large number of free level hydraulic tunnels, with a non-reinforced concrete lining (Figure 14.14).

An inverted vault has been installed on the bottom of this section. This provides strength to withstand the possible swelling of certain types of soil

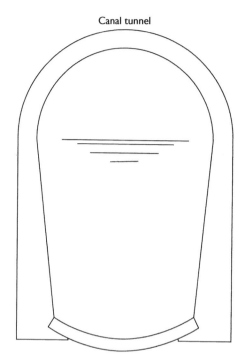

Canal tunnel

Figure 14.14 Trapezoidal cross section with two vaults for tunnels.

due to filtrations from the canal. There have been cases where the ground contained marls and water leaking from the canal caused the ground to swell to such an extent that the flat bottom was fractured.

Evidently, the flat bottom is far from appearing like the ground's thrust line, which is a desirable condition for the survival of all tunnel elements with non-reinforced concrete linings that cannot withstand bending. It must not be forgotten that in tunnels, as in deep trenches, there is a tendency for the ground inside the tunnel to creep and ascend from the bottom due to the weight of the columns located on either side of the tunnel. This is why it is so important to construct the bottom in the form of an inverted vault.

All thrust lines have shapes with curves that have greater curvature with increasing applied loads per unit of length.

We can therefore state that the tunnel's lateral walls, with flat extrados and intrados, which also form the vertical force support walls (Figure 14.15), so often successfully adopted, are not the best solution from the strength point of view and should have a certain curve. This would allow them to more easily withstand the ground's active lateral thrust.

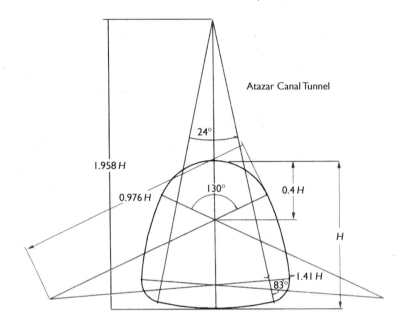

Figure 14.15 Cross section of the Atazar Canal for water supply.

However, the upper vault has to support the vertical weight of the uncompressed ground above, which is normally far greater than the lateral thrust acting on the lateral walls and should therefore have, in accordance with the previous observation, greater curvature.

This line of reasoning leads to a cross section which is something similar to that shown in Figure (14.15).

In this situation, the vault has a greater curvature than the side walls. Since the width in the bottom zone is maximum, the load acting on the ground per unit length is less than that on the vault and hence the curvature can be much less.

The section between the bottom and the side walls is completed with circles of agreement.

From the mechanical stability point of view, this section has successfully withstood the test of time in the Atazar Canal over many years and its section is shown in Figure 14.16.

However, using the same reasoning for a section with a shape similar to the ground's thrust line, the shape shown in Figure 14.17 could also be reached.

This also adopts less curvature for the side walls than for the vault, but because its greater inclination reduces the bottom width, the vertical loads

Figure 14.16 View of the Atazar Canal for water supply to Madrid.

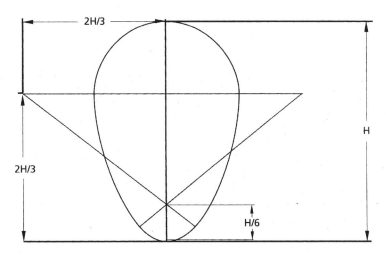

Figure 14.17 Egg-shaped cross section for hydraulic tunnels.

acting on this per unit length are much greater than that of the previous section and so a large curvature has to be used.

For a very long time, this section called "egg-shaped section" was very often employed in sewers, and provides the advantage that, since it is high, it can be made accessible, even in the case where the total useful area required for the flow of water is small.

Without ignoring the classic section of Figure 14.14, the sections shown in Figures 14.15 and 14.17 should be taken into consideration in any comparative studies.

Tunnels are evidently not the subject of this book; however we should at least state that the constructional method to be adopted must influence the chosen section. We should also mention the possibility of excavating the tunnel with tunnelling machines that impose certain dimensions on tunnel size, which the design engineer should adopt.

We would also like to point out the development of the so-called Austrian method which, instead of employing a lining that is able to withstand the ground thrust, makes use of a ring of the ground itself, which possesses tensile strength because it is reinforced with bolts and also adding a gunite or concrete layer sprayed onto the inner face, which will impede or prevent weathering of the ground.

The basic aim of the Austrian method is a vault that is able to withstand the ground thrust, which consists not of a concrete lining, but instead a ring of ground reinforced with bolts. A stress survey must be carried out by an expert in tunnels, possibly with the assistance of a computer application. The truth of the matter is that the ground is more or less decompressed and this result is greater the closer it is to the strengthening ring or vault, even when the ground layers are peeled away like onion skins in the area close to the tunnel.

The final situation of balance is achieved because part of the uncompressed ground acts on the vault and the upper part of the mass maintains itself because it is not altered as much. This leads to the simplified diagram shown in Figure 14.18, which can also be seen in the book (90) and which is based on the Protodiakonov Theory which, although is very old, is still very illustrative. That part of the ground acting on the vault is defined as that under a parabola, the axis height of which, according to Protodyakonov, is as follows:

$$H = b/2f$$

where

   b is the parabola base width
   H is the parabola height over the vault keystone
   f is a coefficient known as the strength factor, with the following values which depend on the class of ground.

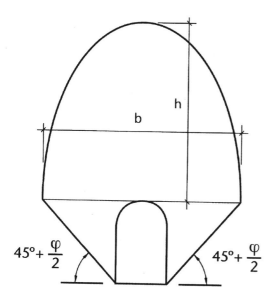

$45° + \dfrac{\varphi}{2}$ $45° + \dfrac{\varphi}{2}$

Figure 14.18 Protodiakonov theory of ground loads.

Ground	Strength factor
Quartzite, basalt and other sound solid rocks	20
Sound granite and very strong sandstone and limestone	15
Granite. Very strong sandstone and limestone. Sound compact agglomerate	10
Limestone. Weathered granite. Sandstone	8
Normal sandstone	6
Slate sandstone	5
Clay slate. Weaker sandstone and limestone. Loose agglomerate	4
Various types of slate. Dense marl	3
Loose shale and very loose sandstone	2
Normal marl. Cemented gravel	2
Compressed gravel soil. Cracked slate	1.5
Hard clay	1.5
Dense clay	1.0
Gravel	0.8
Soil with vegetation. Wet sand	0.6
Sand and fine gravel	0.5
Silt soils	0.3

The active ground thrust acts on the side walls, the governing laws of which are used to define the parabola base.

If the ground is very sound, then the parabola height is zero and the tunnel does not require a very strong vault.

From the Protodiakonov Theory, it can be seen that stability problems grow rapidly with an increase in tunnel width. One consequence of this is that whatever the stability of the ground, the construction should begin with a small heading that is sufficiently lined to provide the required stability as the excavated section dimensions are increased.

The various tunnel excavation methods vary depending on the section widening process, always combining the excavation process with the lining of the areas that provide the necessary stability and which are conserved in the final section.

The heading should be as small as required by the ground instability, but for construction reasons, it must allow a wagon to pass, which means a minimum section of $3\,m^2$.

By taking all this into consideration, we are then in a position to improve or optimize or even calculate (with specialist assistance) the shape of the transverse section for the various tunnels from a mechanical strength point of view, and it is also necessary to examine the hydraulic operation (Section 14.3.3).

As with all tunnels, it is essential that the concreting operations of the vault are correctly carried out, without leaving any cavities between the ground and the lining. With this in mind, it is highly recommended that the concreting operations are carried out using pumps followed by a series of cement mortar injections that will fill in even the smallest cracks that may have been left without concrete.

With respect to the plan layout, canal tunnels are usually designed along a straight line between the optimum entrance and the exit points. There are two very significant exceptions to this. An infrequent one occurs when specific problems reveal themselves at determined points, which should then be avoided. The other involves very long routes, which drastically increase tunnel costs, partly because excavation products have to be transported further and partly because of an increase in execution times with consequent investment capital interests. In such cases, it is normal to replace the straight section by another in a polygonal line that enables the vertex to be located at a point close to the hillside, which is favourable for the location of an entrance adit. The tunnel is thus divided into two, each of which can be constructed from both ends, resulting in a lessening of the described inconveniences.

In any case, a prior, highly accurate topographic survey is necessary, which should include precise triangulation and levelling in order to have all the portals correctly linked up. Accurate topographic work should continue throughout the entire construction in order to ensure all alignments and slopes are correctly followed.

14.3.2   Pressurized galleries

Pressurized tunnels are not described to any great depth in this book, because it is considered that these, in the same way as the penstocks at hydroelectric power plants, have their very own peculiarities. We will limit ourselves to stating that pressurized galleries usually have a circular section in order to better withstand the internal pressure and also, no mater how small the operating pressure is, they must have a reinforced concrete or even a metal sheet lining.

In this type of piping, the tunnel has to be able to withstand two different and opposite types of stress: the rock thrust and the internal water pressure. The latter may disappear when the tunnel is not in service. The rock thrust is held back by the concrete lining and the internal pressure by the reinforcement or metal sheet lining.

On certain occasions, the reinforcement is not necessary because the ground consists of firm rock. In such situations, consideration is given to the rule whereby the ground thickness above the tunnel is equal to or greater than the internal water pressure. Since the relative density of rock with respect to water is around 2.5, the ground pressure will counteract that of the water, with a corresponding safety factor.

In a first approach to dimensioning the amount of reinforcement steel or metal sheet lining required to withstand the internal water pressure, it is normal to calculate the section so that the steel is working at its elastic limit with respect to this stress, and trusting the rock to withstand the rest of internal pressure. The reinforcement steel is usually distributed as shown in Figure 14.19, which refers to the Talave Tunnel on the Transvase Tajo-Segura Canal.

All this is associated with classic criteria for a first approximation. Modern calculations performed with a computer using the finite elements method produce greatly improved approximations of the actually required dimensions.

14.3.3   Hydraulic operation of free level tunnels

The existence of a vault in free level hydraulic tunnels produces a substantial difference when compared to open canals.

If we assume that the free level of the internal water ascends close to the keystone, we will find that, from a certain water height, each increase in level involves an increase of interior width, and hence an increase in

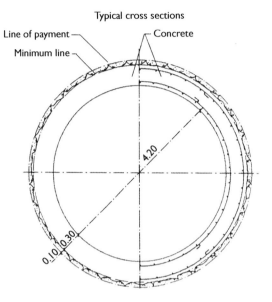

Figure 14.19 Reinforced concrete lining of the Transvase Tajo-Segura Canal.

useful area, which is increasingly smaller; while on the other hand, the wet perimeter undergoes great increases because its mild inclination with the horizontal causes it to increase much faster.

The result is that when the water height is raised, the hydraulic radius, which is the area divided by the perimeter, worsens, which will tend to reduce the velocity.

Although the useful area has been increased, which is a favourable effect for increased flow rate, it will eventually cause the flow transport capacity to decrease.

This may be seen in Figures 14.20, 14.21 and 14.22, in which the flow rate laws are represented as function of the water height for circular, Atazar and egg-shaped tunnel sections respectively.

Figure 14.20 shows that for a circular section, a maximum transportable flow rate is produced for a water height equal to approximately 88% of the diameter and, for a full section, the transport capacity is some 8% less than the maximum. The flow rate corresponding to the full section is also produced for a water height equal to 80% of the maximum.

Figure 14.21 shows that, for an Atazar type of section, a maximum transportable flow rate is produced for a water height equal to approximately 95% of the diameter and, for a full section, the transport capacity is some 5% less than the maximum.

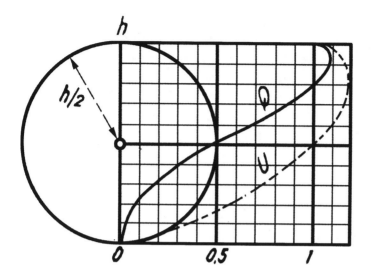

Figure 14.20 Flow rate law in circular tunnel sections.

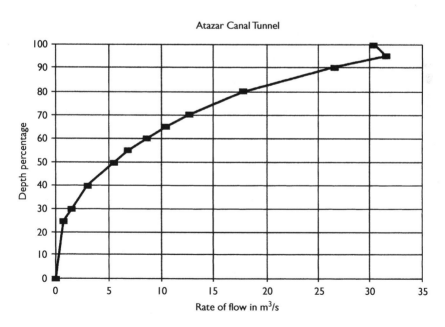

Figure 14.21 Flow rate law in Atazar Tunnel section.

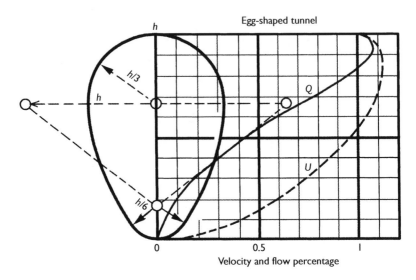

Figure 14.22 Flow rate law in egg-shaped tunnel section.

Figure 14.22 shows that, for an egg-shaped section, a maximum transportable flow rate is produced for a water height equal to approximately 93% of the diameter and, for a full section, the transport capacity is some 8% less than the maximum. The flow rate corresponding to the full section is also produced for a water height equal to approximately 83% of the maximum.

If these curves are used to establish the necessary water height to transport a flow rate that is higher than that for the full section, we will find that there are two possible water heights, one above and the other below the maximum flow rate.

This means that for this flow rate range, there is a certain potential instability in the canal's hydraulic operation because, unexpectedly and for a variety of reasons, the level may pass from one water height to another, and even attain alternative water heights through an oscillatory phenomenon.

It must be stressed that this possibility may even occur when the two water heights correspond to a slow regime, although it would be much more involved with a fast regime.

There are a large number of canals in existence that have sometimes worked with flow rates falling within this range, which have produced an oscillatory phenomenon leading to unwanted consequences and it is therefore recommended that the tunnel's hydraulic design should ensure that the transported flow rates are always within the range of levels for which each flow rate corresponds to a single level.

It must not be forgotten that any sharp variation in water height inside the tunnel will produce brusque compression or decompression in the air above, which could then increase instability.

If we remember that a freeboard is always included in the design to provide a safety margin with respect to increases in flow rates, we should conclude that the tunnel should be dimensioned hydraulically so that the water height corresponding to the design flow rate, plus the freeboard, lies within the zone of stability. The freeboard inside the tunnel should never be less than 1 m.

We can also conclude that in the choice of a transverse section for the canal, it is recommended to select the one for which the zone of stability is as high as possible.

14.3.4 Impermeabilization problems in hydraulic tunnels

We have already described in Section 2.1 the need to combat water losses. The possibility of water loss is much more probable in a canal than in a tunnel because apart from the fact that the ground where tunnels are constructed are normally much more solid and impermeable than those where canals are sited, the geometric filtration slope to the exit is much gentler.

Nevertheless, we have already described the problems produced in a canal due to ground swelling caused by filtration of the transported water.

If, in roadway tunnels, it is necessary to include impermeabilization against the entry of water coming from outside, hydraulic tunnels often require protection against filtration of the transported water when they cross ground that is susceptible to swelling, mainly for stability reasons.

The most dangerous zone, which greatly differs from the rest, is the one lying between the bottom and the side walls in tunnels having a section similar to that shown in Figure 14.14. The contact zone between the bottom and the side walls in these types of canals has to be well impermeabilized.

Other zones, which are also dangerous, are those between concretes of different ages. These are normally found between facing zones having different curvatures, which is usually at the joint between the vault and the side walls in classic tunnels. All these areas and any others that might be considered dangerous should also be impermeabilized.

Transverse cracking may also appear in concrete tunnel linings because of concrete shrinkage or contraction due to drops in temperature, which can evidently lead to potential filtration as was described in Section 4.2.3. As explained here, the separation between joints is inversely proportional to the link force between the ground and the lining. Because this force is very big in tunnels, due to the irregularity generally found in the excavation, the transverse joints would have to be so close that their use is not normally applicable. Shrinkage cracking can therefore only be reduced to a certain extent by adopting suitable precautions during actual construction, choosing low water–cement ratios, very careful aggregate doses and even the use of special cements. Fortunately, temperature oscillations inside the tunnel are small and hence the amount of cracking will also be small.

If the ground crossed by the tunnel is of sufficient quality in terms of not only mechanical strength, but also impermeability, for example quartzite, certain types of granite and grauwacas, etc., assuming that they are not cracked, then the tunnel can be constructed without any lining. However, the excavated rock will have a very large number of protruding sharp edges which will considerably increase rugosity and reducing hydraulic capacity. It is generally recommendable to line the tunnel bottom and side walls to a height equivalent to the maximum design water height plus freeboard, leaving the vault without a lining. The result may then be thought of as a canal inside a tunnel.

If possible leaks could negatively affect the tunnel's mechanical stability, for example by producing uplift pressure during rapid emptying, the design engineer should consider the possibility of installing some form of drainage under the canal bottom, employing the same criteria as those for an open air canal.

14.3.5 Cut and cover tunnels

It sometimes happens that the ground height above the tunnel vault is too small to permit the excavation of a normal tunnel because it would not be stable. In such a situation, it could work out more economic to carry out the entire excavation in the open air, even when the actual excavation is very deep.

The excavation of such a large trench is not stable over the long term unless the volume of the moved earth is greatly increased in order to form less steep slopes and it is therefore usually more economical to construct a vault to cover the canal and cover this with the excavated soil. This produces what is known as a cut and cover tunnel.

The transverse section consists of two side walls which, in addition to withstanding the exterior ground thrust, support the vault as abutments and these are joined by the bottom, which is generally in the form of an inverted vault. The vault has to be able to withstand all the soil that is placed on top since no collaborating strength can be expected from the soil above, even if this is compacted, something that does not occur very frequently.

The upper vault should not be lowered too much because the lateral thrusts would greatly increase and could lead to small rotations in the side walls, which are often founded in damp, unreliable ground, producing a cracked vault. It would be better to employ semi-circular vaults or almost so.

14.4   Bridges over canals

A canal always represents a significant separation between two areas of land, one on each side. This is why either inverted syphons have to be constructed underneath existing roadways or bridges that cross over the canal.

As a general rule, existing roadways and lanes should be conserved. However, at times there is heavy agricultural and even industrial development that forces an increase in the number of crossing works.

Inverted syphons introduce head losses in the canal, which have to be taken into careful consideration prior to a well-thought-out design. On the other hand, bridges are easier to include in the plan, even when the canal is actually in service. Because of this, they deserve a few words.

The structure does not usually present any problems, and straight beam bridges are normally employed, for which there are standardized bridge collections with the necessary spans and loads. At times, if the bridge is included right from the beginning, the canal transverse section is replaced by a rectangular one, with self-standing side walls. These can then be easily employed as abutments, although certain North American standards advise against this, and since this arrangement can considerably reduce the span, a certain economic saving can be obtained (Figure 14.23). However, it should be obvious that certain transitions will have to be constructed between the canal's rectangular and trapezoid sections.

Figure 14.23 Bridge over a canal, using their standing side walls as abutments.

Figure 14.24 Lateral roadway of the Trasvase Tajo-Segura Canal (Spain).

Figure 14.25 "National Water Carrier" in Israel.

It is important to point out that the bridge must not interfere with the space intended for the freeboard in order not to obstruct the passage of water at difficult times. Since any existing beams can have a height of certain significance, the bridge level will be higher than the roadway at the side of the canal, with which it is to equal in order for the line of communications to coincide. This generally means that ramps have to be included in the canal service road. In addition to this, it is necessary to include a plan curve to join the service road and the bridge, which involves a very careful study of the junction, as can be seen in the Trasvase Tajo-Segura Transfer Canal (Figure 14.24). There are occasions, however, when the roadway is designed quite a distance away from the canal in order to have sufficient space for stocks of materials and repair works, etc., as can be seen in Figure 14.25 which shows the National Water Carrier, a canal that crosses Israel. In such situations, the junction between the roadway and the bridge over the canal is much simpler.

Chapter 15

Flow rate measurement works

15.1 Flow rate measurement: Use of weirs

In general, the operation of a canal requires information about the circulating flow rates at various points. In free level canals, this is generally achieved through the use of special concrete works known as measurement structures, which make use of the hydraulic properties of weirs, with thin or thick crest as applicable, in order to establish the circulating flow rate in function of the water energy height upstream. It is only necessary to know the measurement structure characteristics, which will then provide the flow rate formula as a function of height. By entering this into the formula, we will be able to obtain the flow rate.

In all weirs, the water flows over a wall that includes a notch, which exactly delimits the width of the spilling water surface.

There are three possible types of weirs. Some have thin crest (Figure 15.1) and others thick ones (Figure 15.2) and within the latter type there are other special ones, such as the Creager and Ogee, etc., which possess the special feature of being the ones that discharge most.

Thin crested weirs are commonly named "sharp crested weirs".

All weirs obey the following formula:

$$Q = c \cdot b \cdot h^{3/2}$$

Where,

Q is the flow rate in m^3/s
c is a coefficient that depends on the weir type and characteristics
b is the width of the weir notch in metres
h is the water energy height above the weir crest, in metres.

Many authors replace coefficient c with coefficient m, which is given by the following expression:

$$c = m(2 \cdot g)^{0.5}$$

Coefficient c is directly replaceable by coefficient m.

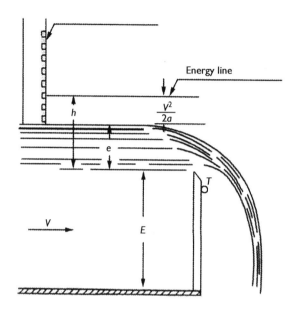

Figure 15.1 Sharp crested weir.

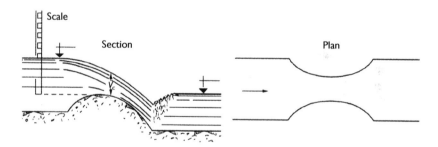

Figure 15.2 Thick crested weir.

Weirs are employed to establish the water flow rate circulating through a canal, based on entering a value *h* into the formula and obtaining the flow rate *Q*.

All weirs have a threshold, which is also known as a crest, over which the water spills and produces a critical speed at a certain point. This is known as the control section. Downstream from this point, the speed is supercritical because, as the water falls, it gains acceleration. On the other hand, the water arriving at the weir does so at slow speed. The energy height should be measured at a certain distance upstream, then this and the specific discharge formula are used together to calculate the flow rate.

Canal or canal network operation requires verification that the flow rates at certain points are actually those necessary, for example, at the beginnings of derived canals, and these same points are where the measuring structures are located.

15.2   Sharp cresterd weir measuring structures

Circulating flow rates can be established with great accuracy with thin-rested weir measuring structures when the water energy height is known.

They consist of a horizontal metal sheet onto which the water spills and forms the fall threshold. The spill width is laterally limited in order to permit aeration of the water surface. If this does not occur, an aeration tube will be required to prevent the water surface sticking to the wall, which would then vary the discharge formula (72) and (46).

The advantage of these measuring structures over those with a thick-walled threshold, which are described in Section 15.4 is that with respect to hydraulic effects, all the metal sheets are equally thin and therefore all the weirs have the same coefficient. On the other hand, in the case of thick-walled weirs, this coefficient can vary significantly depending on the actual thickness.

The problem that has to be resolved in all weirs is to somehow introduce the height of the energy line over the weir threshold. The energy line is an imaginary line, parallel to the water level and located above it at a distance of $v^2/2 \cdot g$, which is the kinetic energy height, where g is $9.8 \, \text{m/s}^2$.

The difficulty lies in the fact that the approach speed is unknown, because the flow rate is also not known and which is precisely the unknown value we want to measure.

The method that is usually employed in thin-walled weir measuring structures is different to that employed in the thick-walled type. In thin-walled weirs, the measuring structures are normally constructed with a widening at the actual entrance of the weir, generally with an increase in depth also, so that the approach speed is reduced to values at which the kinetic energy of $v^2/2 \cdot g$ can be ignored for all practical purposes and that it can be assumed that the water level and the energy line coincide. All that is required now to directly read the water height, which coincides with the energy line and has to be entered in the formula, is to place a graduated scale, levelled with the threshold height, upstream at a distance where the water level depression produced by the water spilling over the crest is not significant.

Approach velocity v should be so small that the expression $v^2/2 \cdot g$, which gives the energy height, is less than 0.5 cm, a value that we can take as being the smallest appreciable in water circulating in a surface bed. Under these circumstances, we cannot be sure if we are measuring the free level or the energy line.

If we find the kinetic energy value for several speeds, we will find:

$V = 1\,\text{m/s}$, energy height 5 cm
$V = 0.5\,\text{m/s}$, energy height 1 cm
$V = 0.3\,\text{m/s}$, energy height 0.5 cm.

We can thus see that the ideal approach speed is equal to or less than 0.3 m/s, but never greater. In this situation, the water level and total energy lines come closer than the accuracy of our visual reading. This method of operation means that if the speed of water in the canal or ditch to be measured has a speed in excess of 0.3 m/s, it then becomes necessary to construct a trough or chamber at the weir entrance with a water height and width so that for the normal maximum operational flow rate, the speed is 0.3 m/s, or very slightly higher.

As it falls, the water drags along air and mixes with it. The danger is that the amount of air held between the falling water and the weir metal sheet is reduced. Subsequently, the vacuum that forms inside produces suction on the spilling water which then is brought closer to the weir metal sheet, which increases the flow rate and renders invalid the measurement being made. This situation will occur if the spilling water width is adjusted to the two sides of the canal outlet to prevent the lateral replacement of the air. In this case, it becomes necessary to install a perforated ventilation pipe under the water surface to eliminate this problem.

The Bazin formula, which gives the measured flow rate for a thin-walled weir is as follows:

$$Q = 2/3 \cdot \mu \cdot b \cdot h \cdot \sqrt{2 \cdot g \cdot h}$$

where $\mu = (0.6075 + (0.0045/h) \cdot (1 + 0.55 \cdot (h/(h+s))^2)$
here

b is the width of the spilling water
h is the energy height over the crest, which is measured as if it were the water level measured as described
s is the crest height over the trough bottom.

If the value of s is sufficiently large, for example three times the value of h, the value of Q will only increase by 3% over the value it would have with an infinite depth. This means that there is no need to take the effect of depth s into account when, in order to reduce the speed, a trough or chamber has been adopted that is not only wider than the arrival canal, but also deeper.

If the trough has a width of B, greater than the weir width b, there is a lateral contraction that modifies the value of μ still further. The formula according to Bazin should use the coefficient:

$$\mu = \left(0.58 + 0.037\left(\frac{b}{B}\right)^2 + \frac{3.615 - 3\left(\frac{b}{B}\right)^2}{1000h + 1.6}\right) \cdot \left(1 + 0.5\left(\frac{b}{B}\right)^4\left(\frac{h}{h+s}\right)^2\right)$$

The widening of the trough above the value adopted for the actual weir itself requires a correction to the flow rate calculation formula.

The choice of weir width is very important. This value is closely linked to the water jet thickness, as can be seen from the general formula given in Section 15.1. For a maximum forecast flow rate, the greater the spill width, the smaller the water thickness.

The decision made for the water thickness greatly affects the accuracy of the flow rates thought to be circulating. Attention should be given to reading errors in the water height measurement due to irregularities in level, to wind effects and errors caused by the difference between the water level and the person's eye, etc. If we accept that the maximum accuracy in the reading is 0.5 cm, then we will have to verify how this value affects the flow rates of the various possible water surfaces with different thicknesses.

The circulating flow rate formula is a general form as shown in Section 15.1. If E is the true flow rate quotient divided by the erroneous flow that we believe to exist then, because of the fact that the falling water thickness is h instead of h+0.005, we will have:

$$E = (h + 0.005)^{3/2}/(h)^{3/2}$$

By entering different values of h, the following values of E will be obtained:

$h = 0.05$ m	$E = 1.15$
$h = 0.07$ m	$E = 1.11$
$h = 0.09$ m	$E = 1.08$
$h = 0.12$ m	$E = 1.06$
$h = 0.15$ m	$E = 1.05$
$h = 0.17$ m	$E = 1.04$

In order to obtain reading errors that do not affect more than 5% of the flow rate, which is the generally accepted error percentage in engineering, it is necessary to have minimum falling water thickness of 15 cm, because smaller values will produce higher percentage errors.

We can now sum up the design process for a thin-walled weir measuring structure as follows:

1 The range of canal flow rates that are to be measured is chosen.
2 Using the formula given in Section 15.1 and coefficient $c = 1.8$ (as an approximate coefficient) or the corresponding value, width b is chosen for the weir, which will produce a falling water thickness of at least 15 cm for the entire chosen range of flow rates.
3 If, as normally happens, the canal approach speed is greater than 0.3 m/s, then a trough will be required to widen or deepen the canal, or both, upstream from the weir, until a speed of less than 0.3 m/s is achieved. Excessive widening can lead to location problems for the measuring structure and for this reason, the trough is frequently deepened more.
4 The corresponding Bazin formula is then applied, depending on whether the chosen trough is equal to or greater than the notch width, by entering the adopted depth value. The formula is tabulated in order to immediately calculate the circulating flow rate once the water surface height is established over the weir crest using the levelled graduated scale.

As supplementary information, we can also say that the graduated scale has to be positioned at a minimum distance of at least six times the maximum falling water thickness, so that the water surface depression does not affect the reading. In addition, to prevent vortexes forming at the water entrance from the canal to the trough, the transition should have a length that is six times the difference in width between the canal and the trough.

The fall of the spilling water produces oscillations and turbulences downstream from the measuring structure which, if they exceed the crest level, will impact against the spilling water, reducing the discharge capacity and invalidating the flow rate reading. Precautions must be taken to prevent the downstream theoretical level being closer to the crest level than a distance equal to the falling water thickness.

Since the water jet cannot be less than 15 cm, the minimum measuring structure head loss is 30 cm.

This inconvenience, together with the volume occupied by the trough, etc., means that this type of measuring structure is only used with small canals and ditches, thick-walled measuring structures, as described later, being preferred for larger ones.

In addition to this, the trough, the main purpose of which is to reduce the approach speeds to the weir, also greatly favours sedimentation of the solids dragged along by the water. This leads to periodic cleaning costs.

Cleaning operations are unavoidable because a significant depth of sediment will reduce the value of s and as previously seen, this will produce incorrect values for the calculated flow rate.

Assuming that because of the natural canal conditions, or because our recommended design instructions have been followed, that the approach velocity is less than 0.3 m/s, that the water depth under the weir crest of at least six times the water jet thickness and that perfect aeration is taking place, it is possible to employ the simplified thin-walled weir formula of Francis (45), which is as follows:

$$Q = 1.84 \cdot (L - 0.2H) \cdot H^{1.5}$$

This formula, which is in metric units, takes the lateral water surface contraction, produced by the reduction of weir width, into account.

The associated flow rate values are given in the table in Appendix 1, which can be used immediately (45). This table includes very small falling water thickness values, which may be subject to the reading errors that have already been described and which we therefore do not recommend. In the table, flow rate Q is given in m³/s and L and H are in centimetre.

Sharp crested weirs have also been constructed using prefabricated concrete sections (85), to facilitate their implementation in smallholding irrigation systems (Figure 15.3).

An outlet can be seen below the weir, the purpose of which is to carry any deposited sediments downstream.

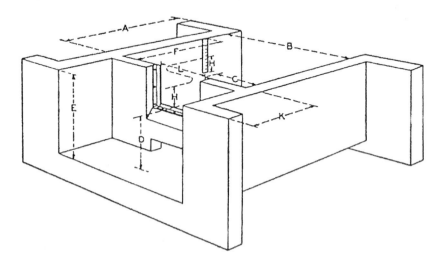

Figure 15.3 Prefabricated sharp crested weirs.

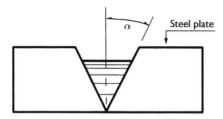

Figure 15.4 Triangular weir.

15.3   Triangular weirs

If the flow rate to be measured is very small, the notch width, when defined using the condition of a minimum 15 cm height, may turn out to be too small. This is when triangular sharp crested weirs come into the picture (Figure 15.4).

The notch has a "V" shape, with the sides defined by an angle of α with the vertical, which is normally 45°, however, it can be whatever size we find most convenient.

Its purpose, even with small spill flow rates, is to achieve a falling water thickness that is sufficiently large, with a minimum of 15 cm for the smallest expected flow rates, in order to eliminate the previously described errors.

In order to calculate the wear formula for these weirs, we assume that the total falling water thickness is divided into vertical strips with a thickness of dx, where x is the distance to one of the ends. The height of one of these elemental water surfaces will be $y = x \cdot \cot g(a)$ and the water spilling over it on application of the formula given in Section 15.1, will be:

$$c \cdot (x \cdot \cot g\, a)^{3/2} \cdot dx$$

The total flow rate spilled is given by the corresponding integral, which is easily calculated by:

$$Q = 4/5 \cdot c \cdot H^{5/2} \cdot (\cot g\, a)^{3/2}$$

Where H is the specific energy line height or that of the water level if they coincide sufficiently, in metres over the weir crest and c is the value calculated in Section 15.2, where a first approximation produces a value of 1.8.

15.4   Thick-walled weir measuring structures

These may be quite different with respect to shape and notch dimensions forming the thick-wall notch. All these produce the critical speed in the spilling water, which is slightly accelerated at the exit. The method of obtaining the critical speed varies depending on the actual type of thick-walled weir. The most frequently employed is that produced by simultaneous narrowing and raising of the crest threshold: however, there are also types that only use one of these procedures.

Since the following canal section is normally operating in slow regime, a jump forms at the exit, which produces what is often called a measuring flume. This is a great advantage because the jump permits partial recovery of the level height lost in the fall so that the head loss in these measuring structures is less than in thin-walled ones.

The discharge factor is not the same when the spilled falling water thickness is small compared to the crest length as when it is large. In the first case, the friction is felt more in the process and the flow rate is less. This is why the formula in Section 15.1 is affected by the Cd coefficient.

The problem of the difference between the water surface level, which can be seen, and the energy height line, which cannot, where the difference in level in relation to the threshold is the value h in the formula, is resolved in this type of measuring structure in a different fashion to those for thin-walled weirs.

If we apply the formula of Section 15.1 using the energy height value h, we will obtain the exact flow rate value. If we employ value h' of the visible water height, in order to have the same flow rate value, we will have to employ a correction factor of Cv in the formula. This will then give us:

$$Q = Cv \cdot Cd \cdot C \cdot b \cdot h'^{3/2}$$

With the measuring structure perfectly defined in the approach section, it is now possible to calculate the value of Cv for each water height h'.

In practice, there is a series of measuring flumes that have been studied by experts that have their own laws for the Cd and Cv coefficients or directly provide flow rate formulas or tables in function of the water level height over that of the throat.

It is essential to construct the measuring structure with the same dimensions as defined by the inventor, then applying the flow rate laws provided by this same person.

Among the most widely used types is that of De Marchi, which is perhaps the first known, that of Parshall, the most widely employed, shown in Figure 15.5, together with that by Boss, the most modern (20).

There are great numbers of publications in existence, which contain standards for various sizes of Parshall measuring structures for a range of flow rates. They are accompanied by detailed tables containing the circulating flow rates for each water height in the canal (28), (45), (85).

They are not strictly necessary because the formula that gives the flow rate in Parshall measuring structures (in m, m³ and s) is:

$$Q = 0.37 \cdot W \cdot (3.28 \cdot H_a)^{1.567 \cdot W^{0.026}}$$

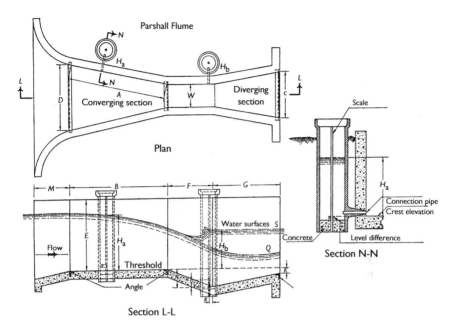

Figure 15.5 Plan and elevation of a Parshall Flume.

where

    W is the throat width

    H_a is the water height level above the threshold and

    Q is the flow rate.

Copies of these tables are provided in Appendix 2.

The table included in Appendix 3 contains all the standardized dimensions for the various sizes of Parshall measuring structures for different throat widths.

The Parshall measuring structure also has certain advantages over the other types. For example, since the throat is not raised above the canal bottom it does not retain solids that could form sediment deposits upstream.

In addition to this, the circulating flow rates are calculated even when there is a certain amount of flooding of the liquid flow at the exit, which permits its use with wider ranges of flow rates. In order to take advantage of this property, it is necessary to have two level scales, one upstream, as with all measuring structures, and another downstream, see Figures 15.5 and 15.6.

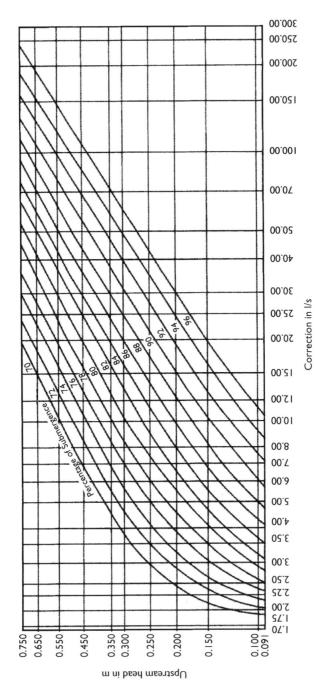

Figure 15.6 Chart for computing the flow rate correction in a 30.5 cm Parshall Flume with submergence.

The result of dividing the downstream water level at the jump exit, above the weir crest by the arrival water level over the same crest, multiplied by 100 is known as the "submergence percentage".

In the case of smaller measuring structures, of 15.2 and 23 cm throat width, the measuring structure reading given by the upstream level is not reduced until the submergence reaches 60%.

For larger measuring structures, the submergence has to be greater than 70% for the reading to be false.

Above these values the Parshall type can be used to measure flow rates based on the two readings from the upstream and the downstream scales, providing the submergence percentage does not exceed 95%.

When submergence exists, the true circulating flow rate is less than that obtained with the general Parshall measuring structure formula. This has to be reduced by a value given in Figure 15.6, where this flow rate appears as a function of the water height upstream from the measuring structure and of the submergence percentage for the case of measuring structure with a 0.30 m throat width.

The upstream water height Ha is entered on the vertical scale, then traced horizontally until it meets the curve defining the corresponding submergence percentage. Tracing the vertical to the horizontal axis enables the flow rate to be read off by which that corresponding to a situation of no submergence has to be reduced.

If the measuring structure is greater, a correction is employed, which consists of multiplying the flow rate to be subtracted by a factor which is given below for each throat width:

Throat width in cm	Factor
30.5	1.0
45.7	1.4
61.0	1.8
91.5	2.4
122.0	3.1
152.5	3.7
183.0	4.3

The flow rate range that can be read using Parshall measuring structures is therefore very large.

Thick-walled measuring structures or flumes usually occupy much less volume than those using a thin weir; they also normally have less head loss and are more economical.

When their sizes do not exceed a certain dimension and, because of the importance of their dimensions being exact, they can be prefabricated in

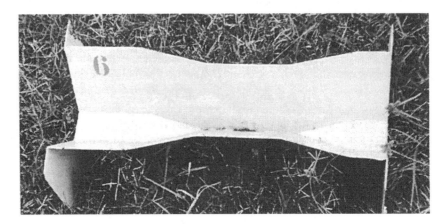

Figure 15.7 Parshall Flume made in PVC – Bella Union (Uruguay).

PVC as shown in Figure 15.7 of the irrigatable area of Calnú en Bella Union in Uruguay.

15.5   Propeller measuring structures

The most exact method of establishing the circulating flow rate through a canal that does not include any form of measuring element is by taking readings with a propeller type meter. However, because of the different water speeds at the various points of the canal's transverse section (see Section 1.5), several readings have to be taken, which are then integrated to provide the results.

The obtained results are not valid for the possible situation given in another moment, since the various water heights will influence the isotach distribution, etc. It is even less feasible to calculate the ratio between the water height at any given moment and the circulating flow rate because of the influence of gates downstream, and other factors affecting the various movements will completely mask the results.

Nevertheless, there are certain points within the canal which are favourable to the installation of mechanical flow rate measuring devices, such as syphon pipes. Although the average water speed is not exactly equal along the pipe axis as at the periphery close to the walls, the speed distribution is very symmetrical around the axis. It is therefore only required to calculate the correction factor that will enable us to pass the axis speed to the average speed throughout the entire pipe.

Speed measuring devices, which may also be flow rate integrators over time, can be located at these points, or other possible favourable ones.

These operate on the basis of a rotating propeller driving a dial meter by means of a gear system.

15.6 Ultrasound measuring structures

This type of device consists of an ultrasound transmitter and a receiver which are able to exchange functions. These are located at each canal bank at two points that do not form part of the same transverse section.

The time taken for the sound wave to travel the space separating the transmitter from the receiver will depend on the distance, which is already known, the speed of sound in water, which is also known, and on the speed of the water, which is the unknown value in this case. If the time taken for the sound to travel from the transmitter to the receiver is known, the exact water speed can be calculated. It is also necessary to know the area of the transverse section occupied by the water, which requires an automatic level indicator. The product of both pieces of information, which the equipment can directly calculate, is the circulating flow rate.

The equipment should be calibrated on installation because the water speed within the transverse section is not that same at all points.

15.7 Flow divisors

Modern irrigation networks are required to provide guarantees of flow rate availability, for example, by means of regulation reservoirs, which provide the necessary flow rates for each plot, but obviously not at the same time, but, instead, in accordance with a specific sequence of irrigation shifts or irrigation turns through the use of certain rules of probability or some other means.

It is in these situations when the measuring and flow rate delivery methods attain maximum significance in order to supply the volume of water that each plot requires.

However, there will be occasions when the irrigation flow rates provided by the canal are not guaranteed and, in accordance with a simple rule, whatever flow rates are available are supplied in proportion with the surface areas of the plots or acquired rights, etc.

Measuring structures can continue to be very useful under these circumstances in order to establish true circulating flow rates, although not as basic delivery or control devices. The structures required in such cases are the flow divisors, which will distribute the available flow rates in proportion to established values, taking them to the necessary watercourses or plots.

These structures consist of a moveable metal blade, which can usually rotate about a vertical axis, although they are also sometimes capable of parallel movement, in order to divide the water flow into two. If this structure is installed in an area of slow water flow circulation, vortexes would

Figure 15.8 Flow divisor in San Martín Canal (Mendoza, Argentina).

form and these, together with the unequal hydraulic radii corresponding to the resulting two water surfaces, would introduce errors that would invalidate the partition. They are, therefore, installed in fast water flow zones, with supercritical water speeds, usually at a weir.

Under these circumstances, the obtained water flows are in proportion to the widths that the spilling water is divided by the blade, without being affected by any turbulence that is produced because it is normal that at supercritical speeds, turbulence is not transmitted upstream. Figure 15.8 shows a flow divisor in the San Martín Canal in Mendoza, Argentina.

Chapter 16

Spillways

16.1 Need for spillways: Types, locations and capacities

Earlier, in Section 10.3, we mentioned the need for the cross section of the canal to be designed with the theoretically necessary depth and a margin of safety that we called freeboard. Several reasons cited there may cause the freeboard to be partially or totally full under abnormal conditions.

In order to cope with the excess circulating water, which occasionally causes overflows, provision should be made in the canal for structures able to empty them automatically into streams or wasteways foreseen for that purpose. This will prevent uncontrolled overflows elsewhere, which could cause damage to crops or to the canal itself.

Such structures are the so-called "lateral spillways" and allow the canal downstream to operate normally, or almost normally.

These may be of two different types. One is a fixed-crest weir, coinciding with the theoretical water surface level, parallel with, or little divergent from, the axis of the canal. When the water rises above the set level, the water spills laterally into a chamber that rids it of its excess energy and is then conducted through an auxiliary course to the place where it is emptied back into the drainage canal (Figure 16.1).

The other type is the syphon spillways. These are made up of a tube with two bends running along the side of the canal at a height at which, when the water level passes a predetermined figure, the syphon fills up and spills all its water, taking advantage of the difference in level between the entrance and the lowest exit, as a result of which both the speed and the flow of water are great, despite the small amount of space (Figure 16.2).

A factor that must be kept in mind in selecting the location of the spillways is the proximity of, or possibility of having available, an evacuation channel, although many times this may require the construction of a canal to carry spillage. The canal's crossings with creeks and streams should always be considered as possible spillway locations.

Irrigation canals and some for the supply of human settlements have the feature of almost always being telescopic, due to the fact that when they feed

Figure 16.1 Lateral fixed crest weir – Sobradinho Canal (Brazil).

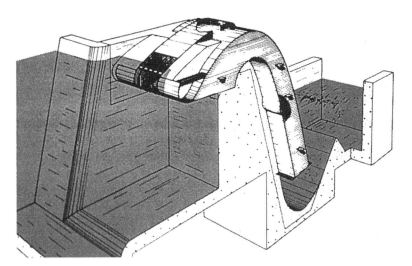

Figure 16.2 Syphon Spillway (Einar).

a smaller canal the theoretical flow diminishes and the following section of the canal is sized with a smaller capacity. Obviously, in such places there may be danger of flooding when, e.g. the total flow reaches them and due to abnormal circumstances (breakdowns, unforeseen change in operational orders, etc.) the derivative inlet is closed. The section of canal downstream from the intake will be inadequate to carry the total flow of water.

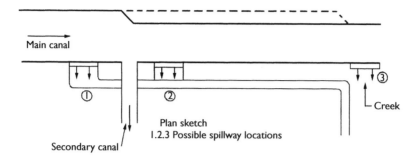

Figure 16.3 Different locations for a spillway.

These places are quite logical points for the placement of spillways (Figure 16.3).

The exact position to be selected may vary depending on the circumstances. Spillway numbered at 1 indicated in Figure 16.3 is inappropriate if the secondary canal is upstream of the wasteway, since an inverted syphon carrying the spilled water would have to be built under the secondary canal. Spillway located at 2 is more appropriate. On the other hand, if the wasteway is upstream, spillway located at 1 would be better.

There is one possibility that should not be ignored. If the evacuating channel is very far downstream, location 2 in the figure could be adopted and an auxiliary canal is built to carry the spilled water to the natural course. But there is also the possibility of placing the spillway downstream, next to the natural course of evacuation and preserve the capacity the canal had down to the new spillway. If the derivative canal takes in no water unexpectedly, the corresponding flow continues to circulate through the main canal up to the spillway, at which point the section narrows. This is solution number 3. Whether or not it is appropriate should be discussed in a comparative study of costs of the possible new canal for carrying the spilled water from location 2 to the natural course on the one hand and the greater expense of the irrigation canal that continues at full capacity up to point 3 on the other hand.

There are no universally accepted criteria for evacuated flows to be adopted in the spillways. Obviously, based on the reasons explained, the spillway should be given a capacity equal to the designed flow of the derivative irrigation canal in the event the latter does not operate. But the possibility that the canal, on reaching the derivative intake, by then had an excess flow due to the freeboard, which must naturally be emptied, may also be considered.

Consideration might be given to the desirability of sizing the spillways to handle the sum total flow of both, but in our opinion this would be too costly given the little likelihood that both situations would coincide.

One reasonable and prudent approach would be to choose the greater of two rates of flow for calculating the spillway: that foreseen in the derivative canal, in the event of non-function, and that which might arrive in excess through the main canal because of the freeboard.

If, for a variety of reasons, an attempt must be made to diminish the estimated flow of the spillway, we must consider two possible cases: that the flow of water through the derivative canal is greater than the flow that may arrive occupying the freeboard or that it is smaller.

The spillway must be able to eliminate whatever flow arrives through the freeboard. But if the flow foreseen for the derivative canal is larger, the situation is more doubtful owing to the fact that there are many canals worldwide that fail to meet the condition of being able to spill the flow foreseen for a derivative canal over a spillway. In order to go so far as to adopt this solution, it is necessary, in our opinion, to foresee the canal's management system very well, the availability of adequate safeguards, etc.

In the United States, the Bureau of Reclamation has built several canals under these conditions. They are canals fed by pumps and, therefore, have little danger of carrying excess flows. Secondly, the gates are managed by centralized automatic systems that guarantee proper operation. This is not normal in many canals.

16.2 Operation of fixed-crest lateral spillways

Fixed-crest lateral spillways operate differently depending on whether the canal operates on a slow or rapid system (at a speed below or above the critical speed).

The types of water surface when overflowing, in both cases, are those shown in Figure 16.4. Most people who have not studied the problem would predict the situation in the second sketch for both cases; and the first sketch would seem to them unlikely.

Nevertheless, the fact that the surface forms in a lateral spillway in a canal are like those indicated can be shown not just in laboratory tests but can also be proven mathematically.

For the study of the spillways it is assumed that the specific energy per unit of water mass is constant. In other words, each unit of water mass (and therefore of volume) that is spilled falls with its specific energy and the volumes that continue to flow through the canal do so at the same rate of specific energy as earlier.

This is the same as saying that the depth of water plus the kinetic energy at entry is equal to the depth plus the kinetic energy at exit. The possible head losses caused by rugosity in the section occupied by the spillway are offset by the slope of the canal. In order for this to be true, one must also assume that the bottom has neither entry to nor exit from the spillway any localized chute or step to cause a change in its height.

Subcritical velocity

Supercritical velocity

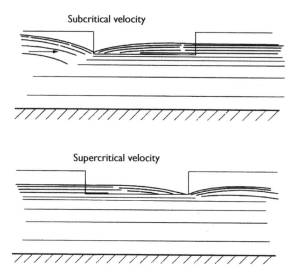

Figure 16.4 Water surface when overflowing lateral spillways.

In studying how the spillway operates, we will call the variable flow of water h, the flow circulating in each cross section of the canal Q, the section of water (different in each section) S, V the speed of the water, V_c the critical speed, A the upper width of the section and g the acceleration due to gravity.

The energy saved will be $B = h + (V^2/2 \cdot g)$ (we call it B in honour of Bernouilli). Its differential will therefore be null, and we can thus write:

$$dB = 0 = dh + \left(\frac{Q \cdot dQ}{g \cdot S^2}\right) - \left(\frac{Q^2 \cdot dS}{g \cdot S^3}\right)$$

As $dS = A \cdot dh$, we can say:

$$\frac{dQ}{dh} = \left(\frac{Q^2 \cdot A}{g \cdot S^3} - 1\right)\frac{g \cdot S^2}{Q}$$

This is the basic equation of a lateral spillway, which we will use to calculate it, but for the moment it will be used to show that the water surfaces types are those indicated. It is

$$\frac{Q^2}{S^2} = V^2$$

as well as

$$\frac{g \cdot S}{A} = g \cdot h = V_c^2$$

and, therefore, we can say

$$\frac{dQ}{dh} = \left(\frac{V^2}{V_c^2} - 1\right)\frac{g \cdot S^2}{Q}$$

If $V < V_c$ (the case of a canal with subcritical velocity), then $(V^2/V_c^2 - 1) < 0$ and $(dQ/dh) < 0$ and when Q diminishes, h increases; therefore the water surface is as indicated in Figure 16.4, upper part.

If $V > V_c$ (the case of a canal with supercritical velocity) then $(V^2/V_c^2 - 1) > 0$ and $(dQ/dh) > 0$ and when Q diminishes, h also diminishes; therefore the water surface is as indicated in Figure 16.4, lower part.

It is thus demonstrated that the surface of the water opposite the weir spillway runs as mentioned earlier.

Subcritical water movement causes a somewhat less useful lateral weir spillway, since if the depth is greater than the height of the crest, the flow that continues downstream is greater than desired. Stated another way, such a spillway is unable to spill all the necessary flow of water in order for the canal to again have just the theoretical flow desired.

What is done is to limit the final excess depth to the exit (and, therefore, the excess flow carried) to a sufficiently small percentage, admitted earlier. For example, 5% more, an amount that is easily absorbed by the *freeboard*.

In order to compute the flow of water spilled (which is necessary in order to select the design length of the spillway), we will follow a procedure to compute the depths of several points of the water surface above the weir crest, at successive and sufficiently close points, starting at a known point, which, if the water movement is subcritical, is the end of the spillway.

In that section we know the depth (that is previously accepted, e.g. the normal maximum depth plus 5%) and the corresponding rate of flow of the canal (calculated using the Manning or similar formula). Based thereon, we gradually calculate, moving upstream, the rates of flow at points fairly close to one another (e.g. 50 cm, or if the spillway is very long, 1 m apart), using a repetitive method in which each water depth is based on the recently obtained earlier depth, at the same time figuring the flow spilled in each elemental stretch using the formula of a normal spillway with a constant water surface height throughout, which is appropriate due to the primary spillway's short length.

Two formulas are used. One is the basic formula of the lateral spillway, already mentioned earlier, which can be stated as follows:

$$dh = \frac{dQ}{\left(\frac{Q^2 \cdot A}{g \cdot S^3} - 1\right) \cdot g \cdot \left(\frac{S^3}{Q}\right)}$$

The other is for the flow spilled by each of the successive elementary weirs with dL width, and given by the known spillway formula, that is

$$dQ = m \cdot (2 \cdot g)^{1/2} \cdot (h - a)^{3/2} \cdot dL \qquad \text{Eq (1)}$$

in which a is the height of the crest of the spillway, also complying with the Bernouilli's equation, in which we call B the total specific energy (sum of the height and that attributable to the speed) and which can be expressed as follows:

$$Q = S \cdot (2 \cdot g \cdot (B - h))^{1/2}$$

Substituting the values for dQ and Q in the basic equation of the spillway, obtained a little earlier, we have

$$dh = \frac{m \cdot (B - h)^{1/2} \cdot (h - a)^{3/2}}{A \cdot (B - h) - S/2} \cdot dL \qquad \text{Eq (2)}$$

The detailed calculation in the case of the subcritical flow is as follows, based on equations (l) and (2):

At the point located upstream from the end of the spillway (at a distance of about 0.50–1 m), the canal will have a flow equal to that estimated for a water depth higher than normal for the maximum flow of, e.g. 5%, plus the water spilled over that distance, in which we can assume that the level is constant and therefore equal to that given by formula (1). The depth will be somewhat less than that at the end point; the difference is given by equation (2).

Based on the data for depth and flow of the point recently estimated, the same procedure gives us the data on the point located at the start of the next stretch upstream and thus successively. Each time we calculate a new stretch, we take the precaution of adding the flows spilled by all the prior stretches. When the sum of spilled flows equals the flow that we want to lighten (the maximum that we assume may pass through the canal less the flow that can circulate with the theoretical depth at the exit, plus the selected increment of, for example, 5%), the spillway has been completed, with a total length equal to the sum of the stretches utilized.

If the spillway were placed in a canal section with supercritical speed, the calculation would be made in a similar fashion but starting with the point in the spillway farther upstream and continuing with the calculation of the points located farther downstream. In this case the water level at the exit may be exactly the same of the weir crest.

Obviously, it is quite easy to write a computer program for the calculation of a fixed crest spillway.

Nevertheless, for a rapid preliminary calculation, the formula given by Engels (44) may be used:

$$Q = 0.414 \cdot (2 \cdot g)^{1/2} \cdot L^{0.833} \cdot (h - a)^{1.666}$$

in which L is the length of the spillway, a the height of the crest and h the maximum admissible depth of water (the theoretical maximum plus 5% or another admissible value).

16.3 Functioning of the lateral syphon-type spillways

The main feature of the lateral syphon-type spillways is that they expel a flow of water that essentially depends on the difference in level h between the canal surface and the free surface at the end of the syphon tube.

The discharge formula is $Q = C \cdot S \cdot (2 \cdot g \cdot h)^{1/2}$, S being the area of the cross section of the syphon tube and C a coefficient. Inasmuch as h is generally a large value (the syphon is located where the wastage channel is low in relation to the canal), large spilled flows are obtained even though section S (and therefore the space occupied by the syphon) is relatively small. The difference with the space occupied by a weir spillways is quite large, justified because its discharge formula depends on how thick the water head above the crest of the weir is (a small value, which also causes very small spill speeds, much smaller than those of the syphon).

The speed of water within the syphon is very high, and therefore the loss of charge is also very high. The purpose of coefficient C is precisely to take into account this effect on the rate of flow.

It is not easy to calculate C, since these values depend on the speed inside the syphon and especially on the more or less elbow shape of the tube. In reality, the formula is: $V = (2 \cdot g \cdot (h - h'))^{1/5}$, in which V is the speed inside the tube and h' is the height due to the loss of charge within the syphon. The h' value is obtainable by a means of false position, using the Manning formula to value the hydraulic gradient inside the tube corresponding to the length of the syphon and adding the losses in the elbows as well. The value for h' thus calculated is used to apply the aforementioned speed formula, calculating a new V that is used as the basis for redoing the calculation.

Nevertheless, a rapid calculation is frequently made by adopting $C = 0.7$. Once the speed V is known, the flow is calculated immediately, since $Q = S \cdot V$.

In most instances, a hydraulic jump is formed at the exit of the syphon even though the wastage canal has a higher level than the exit since the speed inside the canal is usually subcritical. This is very favourable, for it increases the alleviated flow, since h in the formula represents the difference between the level of the canal and that existing at the exit of the syphon, which, if there is a hydraulic jump, is that of the shaft of the tube. Only when the hydraulic jump is submerged it is necessary to adopt as h the drop to the water surface in the wastage course.

Although the initial tendency is to adopt a circular shape for the cross section of the syphon, the rectangular section is usually preferred, for in that way several annexed syphons can be connected in battery shape, facilitating the use of a greater or lesser number of prefabricated elements. A screen is usually placed at the entrance to keep out leaves, branches and pieces of plastic.

The process of feeding the syphon is very important, i.e. the process by which the water that begins to spill freely inside the syphon drags the air from inside and causes it to fill up suddenly, at which time the waste formula cited earlier becomes applicable.

To facilitate this, a sort of small springboard is placed inside the tube (Figure 16.2). In order to exactly determine the canal level at which the feeding occurs, a horizontal slot is placed near the syphon entrance, through which air enters as long as the canal is lower and prevents feeding until the level rises above it and the only way to refill the interior space, which has been left without air, is with a full influx of water. At that point the syphon is fed. This same result is obtainable with elbow tubes that feed the upper angle with air, and enter at the same level as the start of the feeding (Figure 16.5).

One problem with the syphons is that they switch instantaneously from alleviating a virtually zero flow to the maximum flow. This may cause

Figure 16.5 Spillway syphon with fixed feeding level – Big Thomson Canal (Colorado, USA).

brusque changes in the water surface of the canal, causing depressions and accelerations, with rapid emptying of the syphon and the total absence of lighter flows. This in turn may soon cause a repetition of the phenomenon, with at times serious disturbances in the operation of the canal if the flow from the syphon is large.

Should this occur, it would also cause intermittent flows through the evacuation channel, with possible harm such as erosion, etc.

Fortunately, this problem can be resolved if, instead of placing a single syphon, a battery of syphons is adopted, taking care to place the feed slots at different levels, close to one another. The changes in flow of the canal are much smaller (those due to a single syphon unit) and the entry in operation of these elements is successive and gradual, both at the beginning and at the end of the spill.

16.4　Outlets and sluices

The lateral spillways described earlier (whether of the weir or syphon type) operate automatically and require that the water level in the canal be at least the theoretical maximum.

There are occasions when the canal must be emptied (because of the end of the irrigation season, malfunctions or other reasons). In such instances, the spillways are of no use nor can one count on emptying the water through the turnouts, which are frequently too high for that purpose. At that point, specific outlets are required.

Their design poses no problem, since they are mere deep gates, either motorized or manually operated, albeit always by specific command and not automatically.

The main issue is to find a wastage channel. Since the spillways must have this resolved, the usual, recommendable practice is to place spillways and sluice gates together. Obviously, if necessary, should the capacity of the spillways be inadequate, the outlets can help alleviate the flow of water.

They should be designed to carry a large flow of water, limited essentially by the capacity of the wastage channel.

Chapter 17

Height loss works

17.1 Need for height loss works

The purpose of hydroelectric canals is to carry water to a point where pressure pipeline begins and from there to the hydroelectric plant, which takes advantage of the difference in height for power production – the higher the drop the more abundant the power produced. Consequently, such canals attempt not to lose more height than that required by the very gradient of the canal, or for economical reasons.

The same is not true with many canals used to supply population centres or for irrigation. At times, the terrain they cross has differences in height, sometimes brusque and localized and other times affecting longer sections. Normally, in both instances the most economical solution is to give the canal the same slope as the terrain and to match its differences in height. Undoubtedly power is lost but its amount, except in infrequent cases, does not compensate for the construction of small hydroelectric plants.

Two types of works are used to adjust the canal to the differences in level of the terrain. One type are chutes (for fairly long stretches with a steep slope) and another type are drops (for localized slopes).

17.2 Functioning of the chutes

The chutes are sections of canal with steeper gradients than normal, which at times even have speed greater than the critical speed. It is interesting to consider which gradients are the ones at which problems of different types begin to emerge.

In free surface spillways for dams, it is immediately evident that in the upper part of the surface, next to the crest, the water is transparent. A little farther down, when the water has already dropped a certain height and noticeably increased its speed, the water becomes totally white.

The cause is the assembly of abundant amounts of air which are dragged and swallowed by the water circulating at high speed. The whiteness is due to the fact that the water is no longer just water but has been turned into foam, the nature of which is different from primitive water. In effect, the space occupied by a certain amount of water after its mixture with the air is much greater than earlier and, thus, its density is much less.

The speed of water affected by the gathering of air depends on a number of circumstances, but we can attempt to gain an approximate idea.

With a water speed of 14 m/s, the kinetic energy $v^2/2 \cdot g$ is 10 m of water height, i.e. equal to the atmospsheric pressure. When by effect of the friction between the water surface and air the water movement is disturbed, the latter easily turns its kinetic energy into pressure, which causes its inlay within the air mass, the latter having inadequate presssure to prevent this from occurring. This then causes the inclusion of the air. Actually, inasmuch as there are many other factors that influence this (height above sea level, climatology, shape of the water surface, etc.), the mixture with air may occur at other speeds. In Figure 17.1 is seen the Villagordo Chute (Picazo Canal, Spain) with the canal as such visible in the background. The beginning of air emulsion is visible at the end of the rapid. Later on is the stretch where energy is dissipated.

Whether or not air/water emulsion occurs in a canal is very important for the operation and design of the chute. If emulsion is formed, the volume of the fluid is much greater and, therefore, there is danger of overflowing. This then requires greater freeboards. In addition, the friction between the walls of the canal and the circulating fluid is different due to the variation in its physical characteristics of viscosity, etc.

Figure 17.1 Villagordo Chute (Picazo Canal, Spain).

Normally, an attempt is made not to surpass 5 or 6 m/s. At these speeds the Manning or analogous formula is applicable (provided various circumstances do not produce air emulsion). Otherwise, at high rates of speed the Manning (and analogous) formulas are not valid, and the Ehrenberg formula (46) should be used, which we do not attach because it is essentially applicable in dam spillways but not in canals.

The formation of air/water emulsion and the consequent overflow in the absence of a large freeboard is not the only problem that may come up in the chutes. Stationary or non-stationary waves may be formed, caused by rubbing of the water against specific points, curves or defects in the canal, which may cause disturbances or even overflows. This is the reason why, even though the rates of speed are less than 7 m/s (approximately), it is a good idea to adopt more freeboards in the chutes than those indicated for the normal canal sections and which we have cited in Section 10.3. It is recommendable for chutes having supercritical speed to have a larger freeboard than normal. Waves caused in the San Martin Canal (Argentina) by the collision at a rapid rate of speed against a curve and a projecting item may be seen in Figure 17.2.

Figure 17.2 Waves in San Martin Canal (Argentina).

Figure 17.3 Turn out in a chute – San Martin Canal (Argentina).

There is also another series of possible problems that may occur in the chutes. It is highly unlikely that the lateral turnouts (or irrigation turnouts, as the case may be) placed therein operate well. Figure 17.3 shows a turnout in a chute section in the San Martin Canal (Mendoza, Argentina). One can see there the whirlpool formed at the entry due to the supercritical speed of the canal.

In a chute inserted in a canal, the water arrives through the canal at subcritical speed (slower than the critical speed) and continues along the rapid at supracritical speed. At the end of the latter, it should again adopt a normal canal speed, which is subcritical.

If the water enters the chute at the canal's normal slow rate of speed, the latter gradually increases as the water progresses along the chute, which gradually shrinks in size, thus diminishing the depth of water. The cross section, which at first was adequate in the upper part, is excessive in the lower stretches, at an unjustified cost.

The advisable solution is to have the water enter the chute at the same rate of speed as the latter, bearing in mind its cross section and slope, which it preserves throughout, and seeing that the cross section of the chute is constant throughout the entire stretch.

As a result of what has been stated thus far, the chute should have three clearly differentiated parts. The first part is a stretch of accelerating speed, in the upper part, which gives the water the rate of speed appropriate to the chute. The second part is the canal section in chute as such, with a constant rate of flow and cross section. The third is a stilling pool which dissipates the power of the chute and lets the water re-enter the next slow stretch without causing disturbances. We shall now analyse each one of the three parts.

The stretch of acceleration should convert the normal slow speed of arrival of the canal into the supracritical speed of the chute. In order for this to occur, it should pass through a point where the speed is exactly the critical speed, in a section called the "control section". Given the facts available to the reader at the present time, the easiest thing is to take advantage of the measuring flumes, epigraph 15.4, by placing one at the end of the normal canal section. The critical speed is reached in the flume.

Next there must be a drop or change in level that allows the water to increase its speed. If the velocity in the chute is called V_R and V the slow rate of speed of arrival of the canal, the difference in level between canal surfaces and commencement of the chute is obviously:

$(V_R^2 - V^2)/2 \cdot g$ (see Figure 17.4)

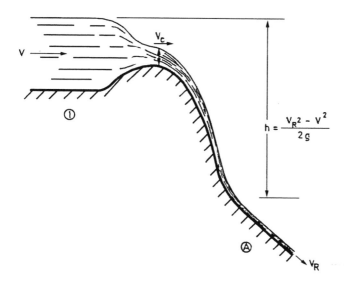

Figure 17.4 First part of a chute.

The floor of the section in which the velocity is accelerating should have a shape appropriate to the speed of the spurt of water. Sufficiently approximate would be to adopt the shape of the curve equal to the parabola described by a mobile thrown horizontally at a velocity of the spurt of water (V_c). Its equation is:

$$x = V_c \cdot (2 \cdot y/g)^{1/2}$$

in which x is horizontal abscissa, V_c the critical rate of speed in the control section, y the ordinates of the curve (taken with a positive sign) and g the acceleration due to gravity. Let us remember that the critical velocity in the control section is:

$$V_c = (g \cdot y_c)^{1/2}$$

y_c being the critical depth.

A countercurve must be placed to connect the prior curve with the slope of the chute being tangent to both. It may be defined as one wishes, with no greater precaution than that it be a mild curve.

The stretch in chute as such is calculated as we indicated earlier by the Manning or similar formula, with the appropriate coefficient of rugosity. As we have indicated, the precaution should be taken to adopt a freeboard larger than that indicated in epigraph 10.3, for example 50% of the theoretical water depth, in order to cope with the undulations that occur at high rates of speed.

The third stretch, destined to decrease the water energy that arrives through the chute, is generally made up of a jump, which is the simplest method for dissipating part of the hydraulic energy. If the depth of the chute is y_1, its velocity v_1, and its Froude number is

$$F = v_1/(g \cdot y_1)^{1/2}$$

the conjugate depth (that which is necessary for the formation of the jump) is:

$$y_2 = y_1 \cdot \frac{((1 + 8 \cdot F_1^2)^{1/2} - 1)}{2}$$

If the water depth in the canal after the chute is not at least equal to y_2, the canal should be deepened through the construction of a pool in order to achieve this value, or better still, as much as 1.1 times y_2 (in order to have a margin of safety). From a totally theoretical standpoint this procedure is not correct, but in practice it is sufficient.

The length of the pool may be $6 \cdot (y_2 - y_1)$ according to laboratory studies conducted (29), (93).

The subdual of the kinetic energy brought by the water is partially dissipated with the whirlpool formed by the jump. The percentage of such subdual is more or less large depending on the Froude number of the incoming water (93), (29). With small Froude numbers, such as those found in the chutes of the canals, the percentage loss of kinetic energy is small, e.g. 20%. Nevertheless, the jump is the easiest method for lessening hydraulic power, and is therefore the one normally used. It further permits a return to the subcritical regime, which is necessary in order to continue with the normal canal downstream. There are possibilities of annulling part of the energy with blocks placed in the floor of the chute and even improving the stilling basin with cross beams (21).

In short, the procedure for designing the chute may be systematized as follows:

One tries the stretch of the chute with a slope that is the optimum, adjusted to that of the terrain. Once a width is chosen for the cross section, the Manning formula is used to calculate the water depth, adopting a freeboard of at least 50% of the depth secured. The speed must coincide with that of the accelerating stretch.

A measuring flume is designed at the end of the stretch of the arrival canal, as well as the acceleration stretch, with the indications given above. If necessary the slope of the chute is corrected and the calculation repeated. Lastly, the formation of the jump at the end is verified, if necessary, by excavating a pool in order to achieve this.

17.3 Drops

These are mere basins to lessen the energy of the water falling from a height situated at a localized point. Often the drop is well defined in the terrain and the slope of the canal is approximately that of the terrain. Other times to the contrary, the slope of the canal should be milder than that of the terrain (to avoid erosion if the canal is unlined, for example), which is done by forming the canal with a series of long stretches, separated by drops, which act like steps to readjust the average slope of the canal to that of the terrain.

The drop is a basin or pool which should meet two conditions, as follows:

1   The first is that it be long enough for the spillage to fall clearly therein.
2   The second is that the volume of the basin, which is given an increase in depth, effectively lessen the energy so that the water does not overflow at the exit or cause erosions downstream.

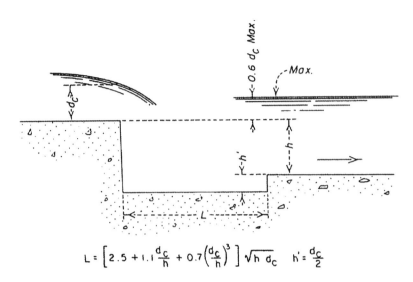

$$L = \left[2.5 + 1.1\frac{d_c}{h} + 0.7\left(\frac{d_c}{h}\right)^3\right]\sqrt{h\ d_c} \qquad h' = \frac{d_c}{2}$$

Figure 17.5 Sketch of a drop (Bureau of Reclamation).

Several theoretical–practical formulas have been studied in order to design the basins for lessening the energy. An interesting one is that proposed by the Bureau of Reclamation, which can be seen in Figure 17.5.
where

L is the length of the basin
d_c is the critical height or water height at fall
h is the jump or difference in level on the floor
h' the increase in depth in the pool $= d_c/2$.

There are other systems for lessening energy that consist in placing reinforced concrete blocks or teeth in the bottom of the pool. There is also the possibility of placing a crossbeam in the pool, which is struck by the flow of water and by the method of impact produces a loss in kinetic energy. These can be seen in the publications of the Bureau of Reclamation (21).

A free fall is assumed in these drops. In the stretch carrying water from upstream, water speeds up and may cause cases of erosion. In the stretch after the drop there may be waves caused by ineffective cushioning of the water, therefore causing the same danger. It is a good idea, therefore, if the channel is unlined, to line the prior and latter stretches, in lengths equal, e.g. to six times the width.

The water surface depression upstream may be eliminated with the so-called "Indiana drops", generally used in unlined drainage channels. These consist in placing a trapezoidal notch at the end of the arriving canal.

Two parameters are required in order to define the notch (for example, the base of the trapezium and the incline of the walls with the vertical incline). These are figured by assimilating the notch to both side-by-side weirs: one triangular and the other rectangular. A system of two equations is posed: one expresses the idea that the flow spilled by the Indiana drops and the arrival line of energy through the canal are equal to those that actually exist for a flow equal to 50% of the maximum (or other rate selected). The other equation is similar but refers to the design capacity and maximum water depth in the canal.

Assuming, for example, a canal in which for the maximum depth y_m the maximum flow should be Q_m, with an arrival speed V_m, and that for an intermediate depth y_i the canal carries a flow Q_i, at a speed equal to V_i.

The energy height for the maximum flow will be $H_m = y_m + (V_m^2/2 \cdot g)$.

Similarly, the energy height for the intermediate flow will be

$$H_i = y_i + (V_i^2/2 \cdot g)$$

Calling a the angle of the slopes of the Indiana drop with the horizontal and b the width of the base of the Indiana drop, the formula for the flow of water spilled, by the sum of the triangular and rectangular weirs will be, in the case of maximum flow and level:

$$Q_m = c \cdot b \cdot (H_m)^{3/2} + (4/5) \cdot (1.75) \cdot H_m^{5/2} \cdot \cotg a^{3/2}$$

For the intermediate flow and level, it will be:

$$Q_i = c \cdot b \cdot (H_i)^{3/2} + (4/5) \cdot (1.75) \cdot H_i^{5/2} \cdot \cotg a^{3/2}$$

The coefficient c is that which corresponds to a rectangular weir. As its crest is in reality the floor of the canal, we assign it a value 1.4, which more or less corresponds to weirs on a thick wall.

From this system of two equations can be deduced the two unknowns a and b.

Chapter 18

Ditches and tertiary canals

18.1 Special characteristics

In irrigation areas, the terms ditches and tertiary canals are sometimes employed.

The actual difference between these two is not clearly explained, although both refer to small canals.

In large irrigation networks, there is normally a single main feeder canal, from which other, smaller canals are derived. These are secondary, which then produce the tertiary, from which water is directly taken to the plot via plot turnouts. Sometimes there also are quaternary canals, etc.

As far as we are concerned, the terms ditches and tertiary canals refer to those watercourses used for the irrigation of smallholdings by means of irrigation outlets. Normally, when these are small, they are directly excavated in the ground, however, on many occasions they are constructed from concrete. On the other hand, when referring to irrigation watercourses, which are not normally connected to plot turnouts, but are intended to feed water to other smaller watercourses, we simply call them "canals".

This is the definition that we shall employ through this book.

This division between the words canal and ditch or tertiary canal therefore involves a certain distinction in operation, as will be described in the Chapter 19. Usually, the aim is to obtain operation with a continuous water flow in the main watercourses, which are the canals, while employing a system of turn irrigation in the direct irrigation watercourses consisting of the ditches or tertiary canals.

The term "turn" or "rotation" irrigation refers to the fact that each direct plot turnout from the ditch or tertiary canal is supplied with a flow rate that the irrigator is able to use during a specific number of hours that will be enough to satisfy this person's irrigation water requirements. Once this time period has expired, the associated plot turnout is closed and another is opened for a time decided by the surface area and the type of crop and so on successively. It should not be necessary to explain that, for operational convenience of the network, rotation and direct irrigation of the plots are carried out using the ditches or tertiary canals, which are therefore the smallest, so that this concept of the term ditch or tertiary canal basically coincides with the term "small canal". This concept will be enlarged upon in Chapter 19, which is dedicated to canal regulation.

The smaller size of the ditches or tertiary canals does not affect either the formulas or the hydraulic operation concepts that have been previously described, and they continue to be completely valid. This also applies to the philosophy of the special works, which are identical to the previous ones, with the only difference imposed by the reduced dimensions. The elements that are conceptually new are the direct irrigation plot turnouts, but these can still be basically treated with the same criteria as for the canal turnouts.

There is a basic difference between the irrigation canals and the ditches or tertiary canals, which is a consequence of their direct plot irrigation purpose. We refer to the fact that the ditches or tertiary canals must have the water surface above the ground that is to be irrigated, assuming that only surface irrigation is being carried out, without spray irrigation pumps or localized irrigation.

This greater water level over the ground should be around a minimum of 20 cm, but if the tertiary canal is not located connecting the higher points of the plot, then when calculating the canal water surface level, it will be necessary to take into account the highest plot point increased by the required difference in level so that the water can reach it using another provisional ditch. When designing the tertiary canal plan layout, it should always be arranged for it to run along the highest ground points, if it is possible.

The water surface levels for the irrigation feed canals are, on the other hand, fixed by the need to supply water to the derived canals with the levels that these actually require.

The other important condition to be fulfilled by the tertiary canals is that in order to comply with their function of irrigating the plots, they must possess a water velocity that will permit a good level of plot turnout operation.

Although special works have been designed to permit the derivation of water to the plots, even with high speeds, it is recommended that limited velocities, between 0.25 and 1.0 m/s, are employed, which should also avoid sedimentation due to very slow speeds. The most usual speeds are around 60 cm/s.

18.2 Tertiary canals constructed onsite

Tertiary canals can be constructed onsite or, as we explained in the chapter on concrete slabs for canals, they can be constructed using prefabricated pieces.

If the tertiary canals are large, then any of the linings and constructive methods that were described for canals can be employed.

If, on the other hand, they are small, and they are to be constructed onsite, it is standard practice to employ sections with self-supporting walls over a concrete bottom. The only difference is that, due to simple construction, vertical slopes are usually adopted, the same in both the interior and the exterior side walls.

This involves a poorer level of stability with respect to tipping than with an inclined exterior slope, as described for canals, but because the dimensions are small, the values of thrust are also small and hence there are not normally any problems associated with stability.

It is normal to adopt a side wall thickness that is equal to 0.3 times its total height, or sometimes one-third. A small exterior backfill is employed that wraps the tertiary canal on both side walls, producing a tipping force that is opposite to that of the water, which is then partly compensated, increasing stability. This is the transversal type of section recommended by the old Tertiary Irrigation Canal Instruction from the Spanish Ministry of Public Works, which was produced in the 1940s and valid for flow rates up to 500 l/s. Obviously, with greater flow rates, it becomes necessary to calculate whether the side walls are stable, even when the canal is empty and, of course, when it is full, verifying that the side walls can withstand the thrust of any backfills.

Concrete tertiary canals constructed onsite, with a rectangular section and self-supporting concrete side walls, should be constructed, as with all types of concrete, using vibrated concrete. The bottom is vibrated with a surface vibrator, and the side walls generally with a poker vibrator that is inserted between the vertical formwork. In accordance with the philosophy described for canal construction with onsite produced concrete, it is essential to employ a formwork or side wall mould system that is both economic and quick to assemble and disassemble since it forms a significant part of the total cost.

This can be achieved using flat metallic formwork that is bolted to rolled sectional steel, which have a sufficient number of holes to obtain the required thicknesses between the side walls (Figure 18.1).

The previously cited Instruction from the Spanish Ministry of Public Works about the transformation of irrigatable zone (as yet unpublished) requires the use of three sizes of aggregates, mixed together in order to achieve the Bolomey Curve or similar. The maximum aggregate size should be 25% of the side wall thickness as is usual in concrete construction, except thin canal linings, for which a third of the thickness is normal.

The normal dose rate of cement is $200 \, kg/m^3$, or perhaps a little more. The water–cement ratio depends on the vibrator power, but should not exceed 0.55. In well-executed works, consideration should be given to the possibility of reducing the proportion of water, including the use of air-entraining agents in order to facilitate concrete compaction.

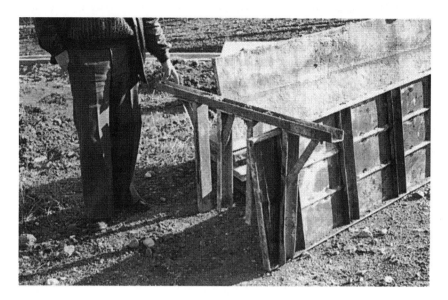

Figure 18.1 Formwork for small canals and ditches.

As with all concrete, a correct level of curing is essential, which will normally require frequent watering of the surface, even before removing the formwork. This operation, as we have already said in connection with canals, and for the purpose of obtaining better economy in formwork investment, should be made just as soon as the concrete has attained sufficient strength. In general, the side wall formwork may be removed 24 hours after pouring in the case of classic formwork.

Traditionally in Spain, the Bazin Formula, with a rugosity factor of 0.3, has been used for the hydraulic calculation of concrete tertiary canals constructed on site, which will provide a safety margin against concrete ageing and silt accumulations in the tertiary canals. There are many books that employ a Bazin coefficient of 0.16 for concrete tertiary canals constructed on site.

18.3   Precast concrete tertiary irrigation canals

Tertiary canals can also be constructed with prefabricated concrete sections. Although they are also sometimes constructed with small prefabricated pieces, for example, those resulting from dividing the tertiary canal into short transverse sections, which are later joined using round longitudinal steel bars forming beams. However, the more general method is to produce complete sections of tertiary canals, which are then supported on piers of a suitable height.

The length of each piece must comply with current legislation with respect to highway transport because the sections have to be transported by lorry from the construction plant to the installation site.

The tertiary canal cross section can be very different depending on the manufacturer and site characteristics, as can be seen from the examples shown in Figure 18.2.

Apart from complying with the associated hydraulic conditions, the transverse section also has to meet all practical conditions. The most important of which is that the external surface of one section should be able to fit into the internal surface of another because they both have the same shape; this means that many sections may be stacked one inside the other. They can therefore be stacked high using a relatively small amount of space, which can be a definite advantage if it is necessary to wait for a certain period of time between section construction and final location. It is also an advantage for lorry transport since the correctly loaded pieces will occupy little space.

Prefabricated tertiary canals are also recommended for large areas of irrigatable land, which will normally involve long lengths of canals that have to be constructed in a short time. The great speed of construction provided by the use of prefabricated lengths makes this method highly recommendable.

This method is also recommended for in areas with two well-differentiated seasons, one wet and the other dry. A very large number of prefabricated sections can be constructed and stored during the wet season and then rapidly installed during the dry season.

The advantage of prefabricated tertiary canals is irrefutable in areas containing gypsum, where normal concretes would be very quickly decomposed by the chemical action of the sulphates. Since prefabricated tertiary canals are mounted on piers, they do not come into contact with the gypsum, only the foundation for the piers are affected and these are quite easily dealt with, for example, by constructing them using stone rubble work and lime mortar or sulphate-resistant cement concrete.

Types of precast tertiary canals

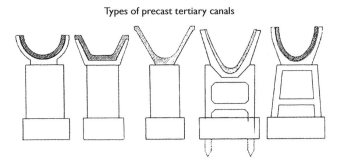

Figure 18.2 Cross sections of prefabricated tertiary canals.

One other very important advantage of the prefabrication method is the better quality concrete.

The employees are able to work under cover, protected from both rain and sun, which results in improved performance and quality, whereas onsite construction requires them to work in bright sun. Moreover, at the prefabrication plant, the employed amounts of aggregate, cement and water can be more closely controlled, in addition to being able to carry out more frequent strength tests on the concrete because the required specialist personnel are available together with test piece fracture equipment, etc.

Since each prefabricated section of tertiary canal is going to work like a beam, it will have to withstand bending moments and the stresses these produce, and so the canal sections are usually reinforced with round steel bars along the full length of the section, or, instead, they are prestressed for the same purpose. Prefabricated tertiary canals have also occasionally been constructed without any reinforcement or prestressing, trusting that any stress caused by bending moments will be withstood by the concrete alone because of the excellent quality that can be achieved. This method should, however, be carefully examined because, in addition to the bending stresses, there are also those caused by low temperatures that tend to contract the piece. If this contraction is limited by the friction of the tertiary canal against the supports or piers, other stresses are produced, which will add to the previous ones and could lead to cracking of the piece, as shown in Figure 18.3.

Figure 18.3 Cracking of a tertiary canal.

In 1964, the Instituto Eduardo Torroja de la Construcción y del Cemento de Madrid (Eduardo Torroja Institute for Construction and Cement) drew up the "Directrices para la Apreciación Técnica de las Acequias Prefabricadas de Hormigón" (Directives for the Technical Appreciation of Prefabricated Concrete Tertiary Irrigation Canals) which, although they are not mandatory, are highly recommended. Among these is the condition by which each tertiary canal section, working as a beam, should be able to withstand its own weight, together with one-and-a-half times the possible weight of water and a weight of 150 kg as a central point weight, without cracking. This point overload is used to guarantee the level of strength even when two persons are walking along the sides of the tertiary canal in order to avoid walking in the mud. It should also be able to withstand the impact produced by a 50 kg load falling from a height of 1 m, which is intended to cover a situation in which a person is carrying a load of this weight and lets it drop onto the canal.

The maximum aggregate size used for the concrete should be equal to one-third of the prefabricated tertiary canal thickness.

The piers may be constructed on site or with prefabricated pieces, with very precise levelling being performed in both cases.

The joints for tertiary canals can be constructed in a variety of ways as shown in Figure 18.4. In some of them, impermeable bitumen cords are inserted between the tertiary canal and the support piece located on top of the pillar. The free space that is left between two consecutive canal

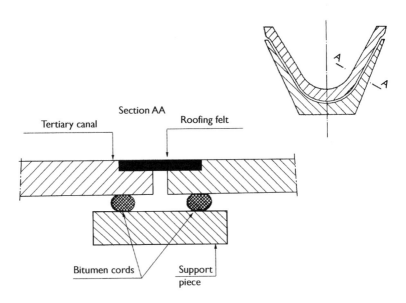

Figure 18.4 Impermeabilization of joints for tertiary canals.

sections as a joint to handle concrete expansion and contraction can be filled with mastic or materials described in Chapter 4. Roofing felt, applied hot over the two sections of the tertiary canal, could also be employed, see Figure 18.4.

In order to avoid the support element, on which two consecutive tertiary canals sections rest, a joint may be used, to which the bitumen felt and impermeable cord are added.

There are also prefabricated tertiary canals that include a hood or bell which is widening at one end that allows a following section to be inserted. Impermeabilization is achieved by means of a bitumen cord inserted between two contiguous pieces. The free space remaining next to the water could also be filled using mortar or putty. However, it should be remembered that if putty is employed, it can then work as both a contraction and a dilation joint, whereas if mortar is employed, this will prevent the two sections from moving towards each other and hence, will only form a contraction joint.

Cost savings with respect to moulds or formwork is important. This is why, at times, if there are tertiary canal sizes that are only going to be used in short length sections, it may be a good idea from an economic point of view to make use only of the larger size, thus avoiding the need for a range of sizes that would be difficult to amortize.

Prefabricated tertiary canals use smaller freeboards that those usually chosen for onsite constructed ones, for which we stated a minimum of 15 cm in Section 10.3. A minimum of 8–10 cm is normally the selected value. This, however, will require excellent levelling of supports, together with assurances that their foundations are not going to settle.

The hydraulic calculation can be carried out using any of the usual formulas, but the associated rugosity factors are generally much lower than those for onsite constructed tertiary canals, which is due to the excellent finish of the concrete surfaces. In the case of the Bazin formula, factors in the order of 0.06 have been employed. Nevertheless, it must not be forgotten that tertiary canals can suffer rapid ageing, due to weather conditions and the surprising amount of vegetation that can cling to its sides, and hence we are of the opinion that a little restraint should be employed in this sense.

18.4 Tertiary canal plan layout

Tertiary canal plan layouts are normally produced on drawings having a scale of 1/5000 with contour lines at 0.50 m.

However, a scale of 1/2000 has also been widely used or contour lines with a 1 m separation, depending on the actual topography of the site.

Because of the maximum distances that water may travel within the plots in normal surface irrigation, the tertiary canals should be separated from each other by approximately 500 m, with a drain or wasteway for excess water in between.

The plan layout should be designed to follow the ground's natural dividing lines or high areas in order to dominate the surrounding land by gravity. This often means that onsite constructed tertiary canals have to follow tortuous routes.

On the other hand, prefabricated tertiary canals, which are supported on piers, the height of which can be cheaply increased, are able to follow straighter lines, naturally following all high points where plot turnouts are to be installed in order to cover all the land to be irrigated. This is an additional economic advantage for prefabricated tertiary canals because their total length can turn out to be somewhat shorter.

Another difference in the plan layout between canals and tertiary canals is that, since canals are very important and large, they are given the most recommendable layout in order to save on excavation and backfill costs. On the other hand, tertiary canals, which are much more versatile in their layouts, are normally located at the boundaries and, because of this, they often have much more pronounced curves than those of canals.

Among the effects studied as determining factors for the freeboard to be adopted (Section 10.3), that of increased elevation at the curves is stressed because of its importance, and which can be calculated using the Grashoff formula. The actual freeboard that is to be used should be very carefully studied using this procedure.

Tertiary canal longitudinal profiles are usually studied with a horizontal scale equal to that of the plan layout, which is the most convenient. The vertical scale in most cases is 1/100, which will provide sufficient detail of the drops, etc. This means that when the plan layout has been established, it will be necessary to carry out ground levelling in order to more accurately define the grade line levels, assuming that the original drawing was not as accurate.

18.5 The prefabrication of inverted syphons and turnouts

When a prefabrication plant for the construction of canals and tertiary canals is installed in an irrigation area, it should obviously be used as much as possible in order to reduce amortization and other general costs, etc.

In addition to the canals, whether self-supporting or lined, which can be produced, the inverted syphons can also be easily manufactured. Since these consist of pipes and transitions or inlets and outlets, which are very different, thought should be given to producing them separately in a form that is easily installed and coupled.

Concrete manufacturing installations, vibrating and curing operations, together with quality assurance and inspection personnel are all the same as for other types of canal. The only new elements that have to be prepared are the moulds or formwork.

The inverted syphon pipes can be manufactured using mass concrete (no reinforcement) up to a diameter of 500 mm, with larger sizes requiring steel bar reinforcement. This is not normally due to internal pressures, since irrigation network inverted syphons, in general, only have to withstand the difference in level between the free water surface and the pipe, which is low, but instead due to crushing problems caused by the soil used to fill in the trench and moving overloads, etc. However, it is recommended that, whenever possible, they are always produced using reinforced concrete, not only because of their possibilities of damage during transport and installation, but, above all, because of differential ground settling.

Because of the importance of avoiding leaks, it is essential that the pipes be fitted with perfectly impermeable joints, which usually consist of natural or artificial rubber rings.

The purpose of this book is not to provide complete coverage of pipe design and construction, which is, in fact, a quite different speciality and lies in the field of the pipe design and construction engineer. There are numerous publications available covering this subject and it is recommended that a start be made with one providing an overall view (66).

Syphon inlets and outlets or transitions, which can have dimensions that make transport and installation difficult, should be manufactured circular, in the form of vertical cylinders, using the same system as for the piping, but broken down into various rings of reduced height, which are then installed onsite one on top of the other. The associated joints will obviously have to be prepared, and these usually consist of extending a cement-rich mortar, as employed in brick fabric, although other product types can provide greater impermeability. The cylinder floor, which also acts as the foundation for the superposed rings, is constructed using concrete produced onsite (Figure 18.5).

Due to the reduced internal hydraulic pressure to which the inlets and the outlets are subjected, the rings are usually constructed with non-reinforced, vibrated concrete, with the required strength being achieved by employing good quality concrete and ring thickness.

The joint between the transition construction and the tertiary canal on the one hand and the inverted syphon pipe on the other requires two rings with suitable notches. The upper ring, which mates with that of the tertiary canal, is easy to make. The only precaution to be taken is that of constructing the notches with the dimensions required for each tertiary canal size. The tertiary canal is supported on the notch, with an impermeable band between it and the ring, which will also function as an expansion joint which, as explained in the section covering this type of joint, is essential.

The lower ring orifice, where the inverted syphon pipe commences, is much more involved, because its strength with respect to the internal pressure is highly reduced. It must be remembered that a ring is able to withstand

Figure 18.5 Prefabricated syphon inlet.

internal pressure because of its perimeter tensile strength, which is greatly weakened because of the orifice.

The solution is shown in Figure 18.6, where two steel rings are fitted inside the concrete ring, one above and the other below the pipe orifice, together with a series of vertical reinforcement elements, all of which forms the only outlet reinforcement. From a mechanical point of view, this

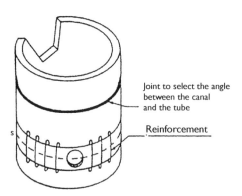

Joint to select the angle
between the canal
and the tube

Reinforcement

Figure 18.6 Syphon outlet reinforcement.

arrangement works like a wine barrel, which consists of wood slats held inside iron rings.

It should be evident that successive rings can be superposed more or less rotated with respect to each other around the vertical axis of the outlet or transition. This means that this type of inlet or outlet is very useful in those situations where the inverted syphon pipe outlet is oblique with respect to the tertiary canal.

Chapter 19

Canal regulation

19.1 Introduction

In order to simplify things when commencing the hydraulic study of canals, it is usually assumed that the speed is constant and that the water surface is always parallel to the bottom, however, this very rarely happens in practice in a permanent fashion.

The inverted syphons, which have already been examined, together with any constructions installed in the canal, will introduce specific head losses that vary with the circulating flow rate. This means they will not coincide with slopes and drops already included in the canal design, which has been calculated for a flow rate that generally coincides with the design maximum. The water surface is then no longer parallel to the bottom, usually for long periods of time and this leads to a regime that we called "varied" in Chapter 1.

Moreover, it is also quite usual to have to vary flow rates or levels artificially over time, or even in the same day, by means of gates that allow more or less flow rate or raising or lowering the levels arriving at the hydraulic regime which we called "variable over time" in Chapter 1.

This is sometimes necessary in order to adjust the flow rates to those required and on other occasions it is necessary to resolve problems in secondary and tertiary canals and turnouts due to a lack of adequate water levels.

All canals suffer from the problem of having a very long response time between introducing a flow rate variation at its beginning until this is available at its destination. This makes canal regulation difficult, where "regulation" is understood as being all the control operations that have to be carried out in order to have the necessary flow rates available at all times along the entire canal. This becomes quite obvious by thinking of the advance warning that is required to modify the feed flow rate at the intake of a canal that is twenty or even fifty kilometres long, with water speeds of between 1 and 1.5 m/s in order to change the consumption at a secondary or tertiary canal located at the other end.

This regulation is significantly more complicated in irrigation canals, which have numerous turnouts, and even more so if the irrigation network is taken into consideration as a whole, consisting of secondary canals leading from the main one which, in turn, feed smaller ones until the actual plot turnouts are reached.

The variation of desired flow rates at a certain point requires the modification of the associated gate for derivation or canal level regulation or both, an operation that will propagate the level variation both upstream and downstream from this point and, obviously, that of the circulating flow rates.

This modification affects the turnouts located downstream and, in general, will also affect the upstream turnouts. An exception occurs in those cases where there is a drop or chute, the supercritical speed of which will guarantee non-propagation of any perturbation in a direction against the current.

This fact leads to a modification of the flow rates in front of the gates, which means that in order to maintain a good level of service, these gates will have to be adjusted, leading to flow rate and level modifications at other gates, so that the requirement for gate adjustment eventually becomes necessary throughout the entire network.

It therefore turns out that the manual operation of a single gate in an irrigation network that includes conventional regulation elements not only requires the theoretical reiterated adjustment of a large number of gates, but also requires this in practice. The associated human costs would be enormous.

However, the worst part of this is the fact that, in spite of everything, success is not guaranteed. It is all too easy to act in an overzealous fashion when opening or closing a gate with the resulting danger of oscillation and increasing instability, together with the very great risk of over-spilling, with the corresponding loss of water or insufficient water flow being supplied.

Such a situation, from an operational point of view, is quite unacceptable. The problem of water loss is combined with poor service, with insufficient water flow at certain times, which would obviously lead to complaints from the irrigators, who would probably refuse to pay the associated water charges if these were based on supplied volume by claiming reduced supplied water.

The great difficulty in regulating irrigation canals is a subject that conditions not only operational methods that are to be employed, but also its entire dimensioning because, as will be seen, the cross section is highly dependent on the regulation system criteria and systems that are to be adopted.

19.2 The traditional way of regulating irrigation canals

The traditional way of resolving this problem, which has been used for centuries, consists in attempting to achieve a situation in which there are no circulating flow rate variations throughput the various canal stretches, or at least these are as small as possible.

This was achieved by assuming the irrigation network to be a canal operating with the same continuous flow day and night, with a water surface parallel to the bottom and permanent water height, which is considered as being the ideal situation. This canal then feeds directly or via secondary canals, which are assumed to operate in the same manner, a series of tertiary irrigation canals (which then supply a series of ditches or direct irrigation canals) with a continuous day and night flow rate.

The concept of equivalent continuous flow rate is closely linked to this form of network operation and is practically the flow rate which, when multiplied by the surface area to be irrigated for each canal section, gives the corresponding circulating flow rate at times of peak consumption.

The plot turnouts are connected to the tertiary canals and are designed to provide each plot with the flow rate it requires for its crops. Because, in the traditional surface or runoff irrigation system, the irrigators have to use a determined flow rate, which was that employed on the plot in a simple, economic fashion, which also was adapted to the ground characteristics, the concept of the farmer's minimum flow came into use.

This is the flow rate that can be handled by the irrigator with the available means and which is supplied to the plot for given number of hours until the required volume of water for irrigation has been supplied.

Since the farmer's minimum flow may be much higher than the product of the equivalent continuous flow rate multiplied by the irrigation surface area, some plots are irrigated while others have to wait their turn.

The tertiary irrigation canals are therefore watercourses that receive a continuous flow rate from the canal, an inescapable condition for avoiding regulation and stability problems, but which supply the various plots with flow rates equal to a farmer's minimum flow, or a multiple thereof if the plot is very large, during a period of time that is equal to or less than the irrigation period. Shifts are established, during which some plots are supplied with water and others are not. The actual situation may be modified over time in order to provide all the plots with a just amount of water. This is normally called rotational system.

It is evident that this concept frees the canal from the regulation problem, but which has been transferred and concentrated in the tertiary canals. The reasoning behind this is that because the tertiary canals are smaller, they are easier to regulate and also because of the fact that any error in operations of one tertiary canal will only have a local effect, without affecting any others.

The various canal sections were dimensioned in this traditional system so that each was able to permanently transport the continuous flow rate required to irrigate the entire surface during the peak period of the year. The flow rates at other times were evidently less, but the transition to these was made gradually in order not to complicate regulation.

One inescapable condition of this method of operation was the continuous irrigation by day and night, a problem that was resolved by actually irrigating at night, with more or less ease, above all, for certain crops such as rice and meadows, although in general with evident poor water usage due to surface irrigation.

When designing a canal, the design engineer adopted, and continues to do so, certain safety factors, to take into account the possible increase in irrigatable surface, the replacement of one crop by another with greater water requirements and the worsening of the lining's rugosity factor over time, etc.

19.3 Modern times problems

Modern times have now imposed such severe conditions that make it quite impossible to continue employing this traditional irrigation system.

First of all, labour costs have greatly increased, above all in developed countries, with very high overtime rates applying at night, and have forced irrigation only at daytime. In view of this situation, in a traditional irrigation network like the one just described, what generally happens is that the circulating flow rates at night, which cannot be regulated and which the irrigators do not use, are lost with clear associated economic losses, together with the inconveniences that the true amount of daily water has diminished because of the spillage.

Secondly, many new irrigation systems have come on the scene, such as aspersion and localized irrigation, etc. which, although they are much more flexible in many ways, have certain requirements with repercussions on the canal, for example the time lost in changing the aspersion heads, etc., which affect the continuity in the water supply.

Thirdly, there is the convenience of readjusting the flow rates in function of unexpected storms and in countries that are not arid or semiarid, and there is the convenience of readjusting the original expected evapotranspiration values, etc., to others closer to the immediate reality. All of this requires greater levels of flexibility in the irrigation network, making it capable of transporting variable flow rates over time, even producing flow rate variations with little advance warning.

This can only be accomplished by forgetting about the old permanent regime with a water surface parallel to the bottom and replacing it with a mixed water transport structure, together with its temporary storage.

This may be achieved either by a physical combination of transport canal and reservoirs to the sides or in the canal itself, which are capable of regulating the unused water, or by a canal in which the levels vary at each point over time, with the storage of the surplus water.

In the first case, there is a problem of head loss between the canal and the reservoir first and then between the reservoir and the canal afterwards, which has to be analysed by the design engineer and absorbed either by the surrounding topography or by pumping. Reservoirs connected to canals will be studied in Section 19.12.

The second case involves two problems: on the one hand, it is necessary to plan for the type of gates to carry out the operation which, because of their complexity, cannot be operated manually and require more or less automatic methods. On the other hand, the canal water heights, or at least the areas of the transverse sections of water, will be larger than those of a traditional canal, because the circulating flow rates during irrigation hours will be greater, since they will have to compensate for non-irrigation times, in addition to representing an additional storage volume as an alternative to that of transport. This requires a study of hydraulic operation based on a variable regime over time, which is more complicated than the traditional method with its stable, permanent regime.

We shall study all these problems separately, beginning with the matter of automatic regulation gates. But we cannot ignore the fact that there are many canals with manually operated gates and that logically many more will be constructed with this type of control system. In view of this, before starting on the automatic gates, we take a general look at manually operated gates.

In all types of gates, the design engineer has to look after not only suitable operation, but also calculation and construction details, etc. In this regard we recommend the German DIN standards (42), (43).

19.4 Manually operated regulation gates

These are the ones used traditionally; however, this does not mean that the techniques employed in their design are always correct.

Figure 19.1 shows a simple model of manual gate proposed in the Irrigation Canal Instruction from the Spanish Ministry of Public Works in 1946. This is constructed from welded or bolted rolled profiles. The fixed part consists of "U"-shaped guides, linked together to form the support for the regulation mechanism. The moving board is sheet metal reinforced with angular profiles to enable it withstand the bending moments produced by the thrust of the water.

It is not recommended for the moving part to occupy all the free interior space of the fixed part of the gate so that inspection, cleaning and repainting operations may be carried out on all surfaces. The water pressure will press the board against the guides, producing an almost leak-tight closure if suitable impermeable elements are fitted. The figure contains some thin bronze plates, one fixed to the fixed part and the other to the moving part. Other metal alloys may be employed providing they are softer than steel. These thin plates are perfectly finished but, because they are constructed using relatively soft metal, they become better adapted and coupled with use.

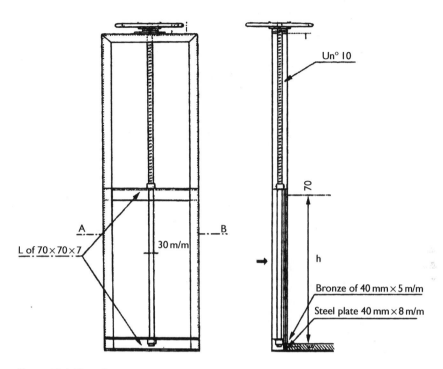

Figure 19.1 Manual gate.

Typical impermeabilization for the bottom is achieved by means of a wooden slat embedded in the bottom concrete against which the lower section of the gate board presses.

Nowadays, the impermeability is obtained for the lateral guides through the use of impermeable material strips, such as artificial rubber or similar screwed to the board, so that they can be easily replaced when necessary. The actual water pressure will force the strips against the canal sides producing the required impermeabilization.

Gate construction as shown in Figure 19.2, where the guides are formed by joining two thin plates, should be avoided at all costs. The small amount of space left between them for board movement will prevent both cleaning and repainting, favouring shoring.

The raising mechanism consists of a steel shaft fixed to the board by means of a cast elevating wheel. The shaft fixed to the board provides a certain amount of rigidity, which can become very useful in preventing the board from bending after shoring when it undergoes violent closing forces caused by unskilled personnel.

When gate size increases, the number of rigidizing horizontal members must also be increased, installing more in between those shown in

Figure 19.2 Gate guides formed by two thin plates.

Figure 19.1. The lower rigidizing members are placed closer together because the water pressure is greater.

Figure 19.3 shows another similar gate, also from the stated Irrigation Canal Instruction, but in this case for submerged orifices. Water is usually lost by the lintel and so impermeabilization is recommended, based on small strips or small cylinders of artificial rubber.

The gate raising mechanism has to overcome its own weight and the friction with the guides, which is the same as the water thrust multiplied by the rugosity factor which, for bronze, is around 0.40 and 0.7 for iron on iron (46). However, some manufacturers use smaller factors.

For widths above 1.5 m, two raising shafts or two chains are required, connected together so that they work at the same time, preventing the board from shoring.

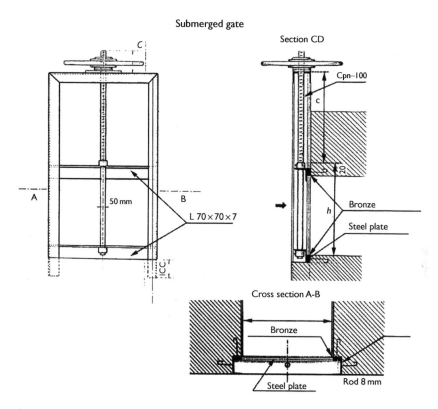

Figure 19.3 Manual gate for submerged orifices.

When the board width is increased still further, the friction of the sides with the guides becomes so great that lifting is difficult. One possible solution is to install a counterweight, normally of concrete (Figure 19.4), hanging from two chains. Wheels can be fitted to the board to reduce friction, which will then allow easy rolling along the guides. It should not be forgotten that the stresses, and therefore the strains also, which are developed inside a cylindrical part and subjected to two diametrically opposite forces, greatly increases with reduced diameter. It is therefore recommended that these bearings have large diameters. We are aware of a case of gates of a size similar to those in the photograph, where the bearings were crushed and had to be replaced by others having a larger diameter.

When the canal is very wide and it has to be divided into various gates, some piles are installed between them. The canal is widened in order not to reduce the effective section and to produce a small head loss, with transitions between the canal's two cross sections (Figure 10.4).

Figure 19.4 Gates with counterweight.

Another ingenious solution for reducing the elevation forces of a flat gate is to add a floatation device to the board, consisting of a hollow, water-tight half-cylinder. The force provided by the floatation device will reduce the weight to be raised.

All these simple, manually operated gates are of the flat board type. The cylindrical board and sector gates have the advantage of radial water thrusts, which pass through the gates' centres of rotation, together with their resultants. It is thus easy to reduce rotation friction and therefore produce soft gate movement.

19.5 Automatic regulation gates

Some automatically operated regulation gates may work by reacting only to the process developed in the area of canal close to them, whereas others operate by taking into account the process of flow rate and level variations at different points all along the canal, even those located at a significant distance.

It is obvious that the second type are able to achieve improved levels of canal operation because they take into account the development of the variable regime phenomenon for the entire canal, partly resolving the canal's slow response time problem. This will require fast electronic data communications of what is occurring at other points along the canal. Its operation is then able to react when faced by different, complicated situations, so that the gate will require a certain level of sophistication that can only be obtained through the use of electrical components.

On the other hand, the gates that only take into account flow rate or level variations occurring close to them are simpler and may require more or less complicated electric control, or even hydrodynamic control.

We are going to briefly look at the various types of these gates, all of which are used to regulate canal levels and flow rates in the following order:

- Automatic hydrodynamically controlled gates responding to nearby variations
- Automatic electric gates responding to nearby variations
- Automatic electric gates responding to distant variations.

These gates will provide near-perfect regulation of the levels through the canal; however, in irrigation canals it is also necessary to automatically provide the required flow rates in the secondary canals and in the tertiary irrigation canals, which will also involve the use of other automatic control gates.

The control gates installed for this derivation purpose can also be either hydraulic or electric, just like those employed in regulation. There will, therefore, be two types of gates used: those for regulation and those for derivation.

The hydraulically operated type for derivation is normally only employed when the regulation gates are automatically operated, so that they are able to make use of their regulated levels.

In order to provide greater understanding of all this material, we shall first describe the derivation gates with hydraulic operating systems, then the regulation gates, whether hydrodynamically or electrically controlled.

19.6 Hydraulically operated derivation gates: Distributors

Regulation gates or systems may not actually maintain the desired water levels, although they come quite close to doing so. Moreover, the water oscillation in front of a turnout, at some distance from the regulation gate is modified by this same distance, by the canal gradient and by the flow rate actually circulating, since the backwater profile curve is different in each case.

The result is that it becomes desirable to have derivation gates that are able to guarantee the supply of a flow rate that is sufficiently constant and equal to that desired, even when the water levels in the feeder watercourse suffer oscillations, within certain limits.

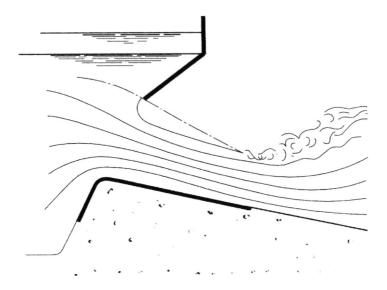

Figure 19.5 Orifice of a derivation gate.

This can be achieved using gates with an orifice, such as that described in Figure 19.5. While the water level does not reach the lintel, the water surface corresponds to that of a free level orifice and is thick. If the canal level increases, the derived flow rate will also increase. But when the water level in the canal exceeds the lintel or baffle, pressure is suddenly applied to the orifice; the stream flow contracts and adopts the typical form of a sub-merged orifice and this tendency compensates in excess the increase in flow produced by the increase in hydraulic head. The result is that the flow rate decreases till a point at which greater increases in level produce large increases in speed through the orifice, without the reduction in the effective stream flow section being appreciable, since it is fully contracted. The result is that the flow rate clearly increases. The corresponding curve is shown in Figure 19.6.

Nevertheless, because of the narrow shape of the derived level/flow rate curve, it is possible, even though the level variation is appreciable, that the flow rates vary only a little, for example by ±5 or ±10% of the average flow rate which, in practice, is a constant flow rate.

This equipment includes an "all or nothing" type of gate. When open, it can guarantee a constant derived flow rate, even when the water level in the feed canal varies between fairly wide limits.

They are generally installed in a series battery formation of several gates having different widths so that the derived flow rate can be "distributed" opening and closing them as applicable, from which they get their name of "distributor" (Figure 19.7). They are also called "modules". This same

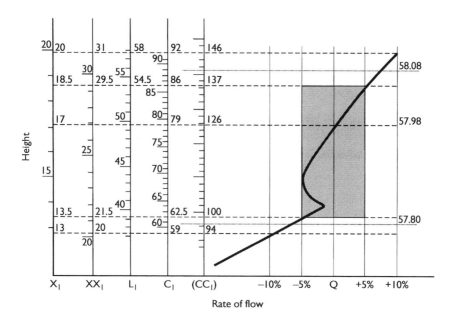

Figure 19.6 Level/flow-rate curve of a distributor.

Figure 19.7 Distributors battery and a regulation gate (Orellana Canal).

figure shows a constant level regulation gate upstream of a type we shall examine later.

The outlet orifice must not be flooded so that the derived flow rate is not altered by the level downstream from the turnout. In addition, in order to achieve the maximum level in the derived watercourse, a hydraulic jump is provoked so that there is a higher level in the tertiary canal that is the equivalent of the conjugated water height level of the outlet stream. This is why these devices are often called "jump distributors".

However, it may happen that the water level oscillation margin in the feeder canal, with which the distributors guarantee a constant derived flow rate, is too large and the described distributor do not work well. This is when the so-called "two-lintel distributor" can be employed.

Two-lintel distributors are normally called "two baffle modules". They are very similar to those already described, but possess an additional possibility when the canal level continues to rise, the water overflows above the distributor and enters a syphon formed by the second lintel that leads to the outlet stream, throttling it and reducing its effective section (Figure 19.8). This will result in an initial flow rate reduction, although, if the level continues to rise, the increased speed due to hydraulic pressure will dominate once again and increase the flow rate.

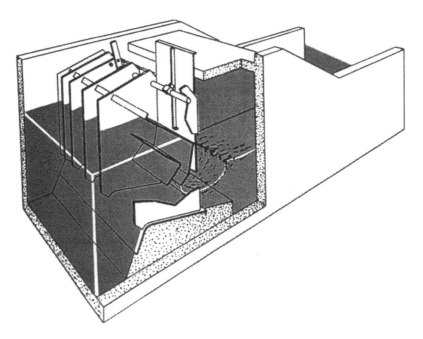

Figure 19.8 Two-baffle distributor.

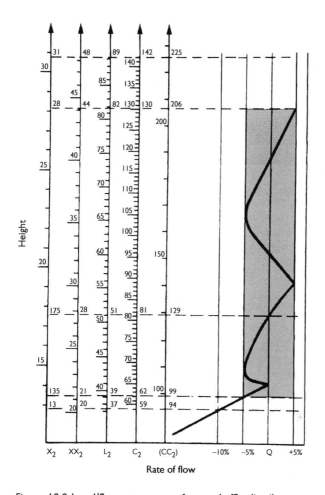

Figure 19.9 Level/flow-rate curve of a two-baffle distributor.

In this way, the derived flow rate/canal level curve once again has an alternative curved and narrow section which permits an even greater level variation while maintaining the flow rate almost constant (Figure 19.9).

19.7 Automatic hydrodynamically controlled regulation gates

These are the best known and widely employed in several countries. Their purpose is to maintain a constant water level at certain points which, according to the type of gates, are located either upstream or downstream from the gates, but always close to them.

This is achieved using a set of counterweights and floats that cause the gate to only be in balance when the water level is that required, otherwise the gate loses balance and either opens or closes in order to establish the desired level.

Each gate regulates the levels for a canal stretch in which the level variations, with respect to that established by the gate, only depend on the canal gradient. This difference in levels will be very small in a short section.

This slight variation of levels in each stretch is made use of by turnout or distributor gates, which only allow the desired flow rates to enter the tertiary and other derived canals.

The maintaining of certain levels within specific margins is, in fact, one way of establishing volumes in a canal that are then used to advantage for regulation purposes.

We are now going to look at two types of regulation gates that can be installed in canals: one type is used to regulate upstream levels and the other downstream levels.

19.7.1 Upstream constant level gates

These are sector type gates that have the advantage of offering little resistance to their rotational movement because all hydrostatic thrust forces pass through the shaft (Figure 19.10).

The gates consist of a front float and two counterweights that are suitably ballasted during installation. One of these is installed at the rear and the other at the top, above the rotating shaft.

The float and two counterweights form a system of forces that tends to rotate the gate, closing it when the level of water upstream is less than the theoretical value and opening it when the level is higher.

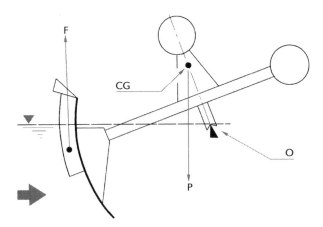

Figure 19.10 Upstream constant level gate.

When the upstream level coincides with the theoretical value, which passes precisely through the rotating shaft, the gate is in a situation of balance that is disturbed by variations in level that force it to readjust its position.

The result is that the gate will allow more or less flow rate to pass underneath it, whichever is required, by opening or closing the amount necessary, in order to always maintain the level established as that desired.

It can be easily demonstrated that, in order for a gate to maintain the level upstream constant for any flow rate less than maximum, it is essential for its centre of gravity to be in a plane perpendicular to the rotating arm, passing through the shaft at a distance equal to $p \cdot l \cdot (R^3 - r^3)/P$, where p is the specific weight of water, P is the total gate weight, l is the float width. R and r are the interior and exterior radii of the float's cylindrical surface.

This conclusion is easy to demonstrate when the following is taken into account: if α is the angle formed between the gate arm and the horizontal in any position, then the gate is subjected to two different moments of force. One is due to the flotation thrust and the other to the total weight. If the centre of gravity is at the indicated location, then both moments are equal and opposite for any angle α for the gate position, provided that the level upstream is that required. If the level should vary, the gate will no longer be in a state of balance and will either open or close more to the passage of water.

In order to achieve the desired gate operation, two adjustments are necessary during installation. The first, with the canal empty, is to ballast the rear counterweight so that the gate is balanced in the horizontal position, in the same way as a weighing balance. This will mean that the centre of gravity runs through the vertical plane passing through the shaft.

The second adjustment is to ballast the counterweight located above the shaft, so that the centre of gravity of the whole assembly does not leave the plane that is perpendicular to the gate arm; however, we do modify its elevation. This operation should be performed with little circulating water until the gate maintains the water level in the required regulation plane. This indicates that the elevation of the centre of gravity is located at the calculated position, which means that the gate will only be in balance when the water level is that required and for any flow rate.

If several gates of this type are installed in the canal, the latter will be divided into a series of sections, as shown in Figure 19.11. Each gate will automatically guarantee a predetermined fixed level just upstream from each one, and the water level along the length of the canal will adopt the form of the corresponding backwater profile curve for each section. However, the two level curves at the ends correspond to the two limit situations of flow rate $= 0$ and flow rate $= Q_{\max}$, respectively. In the first case, the canal consists of a series of horizontal level reservoirs, and in the second the water surface is parallel to the bottom in accordance with the uniform regime in

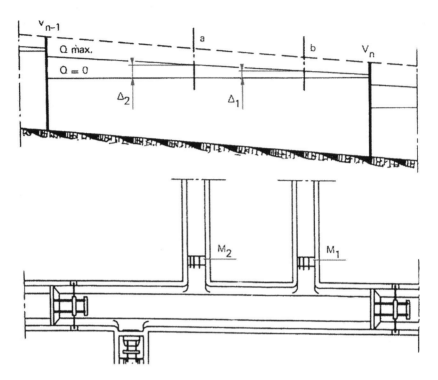

Figure 19.11 Canal regulated with constant upstream level gates.

the canal, although a step will be formed at each gate corresponding to the head loss, an extreme situation that must be suitably covered in the design.

Any derived turnout upstream from a gate will have a water level in front of it between flow rate = 0 and flow rate = Q_{max}. If this variation can be absorbed by a two-baffle distributor, located at the beginning of the derivation, then we can be sure that the associated flow rate is practically that desired. If, when designing the canal, we cannot find either a one- or a two-baffle distributor capable of this, then the regulation gates will have to be sited at other locations, so that there are adequate oscillations in front of each turnout.

A canal designed in this way will provide the planned flow rates at all turnouts. If we operate one of them, either opening or closing it, the constant level gates will automatically operate to conserve these same levels, without any need for manual intervention, and thus maintaining all derived flow rates.

However, this canal has an important drawback. It has to arrange for the total flow rate that enters at its main intake to be equal to the sum of all derived flow rates, which is, in fact, difficult to produce. If it is less, then the gates installed in the first part of the canal would either fully or partially close in order to guarantee its levels and therefore the derived flow rates for the first turnouts, but leaving the end of the canal dry, which would be the part that suffers the lack of water. If, on the other hand, the total feed flow rate at the canal's main intake is excessive, then all the gates would open in order to allow the surplus flow rates to pass through, which would be lost at the other end of the canal. This is one disadvantage of this system, which can be partly rectified by means of a regulating reservoir near the end, before the last turnouts, which must be fitted with distributors that are able to withstand the forecast level variations in the reservoir.

19.7.2 Downstream constant level gates

The previously described disadvantage suggests that the requirement for a canal operating "on demand" in the same way as a water pipe under pressure is able to supply flow rates when the associated taps are opened, but which becomes inactive, without any water losses when they are closed.

This is achieved using automatic gates that guarantee a constant level downstream from their locations. They are the same as previously described, but with the float installed on the downstream side so that they regulate the level in this zone (Figure 19.12).

They also have a counterweight fitted to the gate's main arm. It is very easy to demonstrate that for the gate to be in balance at the required water

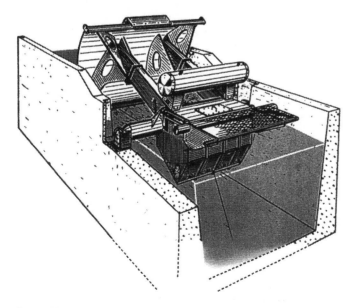

Figure 19.12 Downstream constant level gate.

level, which coincides with the gate shaft, and also for any flow rate less than the maximum, all that is necessary is that the resulting centre of gravity is to be in the plane perpendicular to the gate arm over the rotating shaft and that it passes through its shaft at a distance d, which is given by:

$$p/(3 \cdot P) \cdot I \cdot (R^3 - r^3)$$

where P is the total weight, p is the specific weight of water, I is the float width, R is the external float radius and r is the internal float radius.

This conclusion is easy to demonstrate when the following is taken into account: if α is the angle formed between the gate arm and the horizontal in any position, then the gate is subjected to two different moments of force. The first is due to the thrust of the water on the float, and the second to the total gate weight. For the gate to be in balance, it is essential for both moments to be equal and opposite, a condition which is met for all values of α; in other words, for any degree of gate aperture, if the stated position is met for the centre of gravity and if the level downstream is that established and coincides with the gate shaft.

If, for any reason, the level downstream tends to vary, the gate will no longer be in balance and will either open or close until the level is obtained again.

The counterweight can be adjusted by moving it along the gate arm and then perpendicular to it. Both these operations must be carried out with the desired level downstream from the gate. The first is intended to produce a situation where the centre of gravity is over the rotating shaft and this is performed with almost zero water flow, in other words, with the gate closed. The second operation is carried out with greater flow rate, with the gate almost fully open and is intended to ensure that the distance from the centre of gravity to the rotating shaft is that calculated.

These gates are used to divide the canal into stretches, such as those shown in Figure 19.13. The level of each one can vary from the horizontal position corresponding to zero flow rate, to the water level parallel to the bottom for the theoretical maximum flow rate. The level of oscillation at each turnout is given by this difference, which has to be absorbed by the distributor.

It is also necessary here to include in every gate a step corresponding to the head loss and for the canal to comply with the need to guarantee the required derived flow rates, in the same way as the previous case. However, there is a basic difference: if a specific turnout is opened or closed, the level in the associated canal section will tend to drop or rise, which will cause the gate in the section immediately upstream to tend to open or close. This phenomenon will then be transmitted to the upstream sections until it reaches the canal head gate, which will open or close in order to supply the necessary flow rate. The canal is therefore working as a pressurized pipeline.

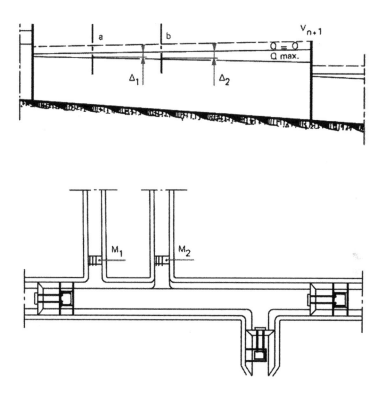

Figure 19.13 Canal regulated with downstream constant level gates.

The head losses will vary in accordance with the gate size and the manufacturer should provide information with respect to these values.

From an operational point of view, there is no doubt that this type of canal provides better service than the previous one. However, it does have its own serious inconvenience. The fact that, for zero flow rate, the canal is made up of a series of horizontal reservoirs means that it has to be designed with a horizontal berm or bank, increasing the water height from upstream to downstream for each section, which obviously involves an increase in the costs of civil works, and perhaps with greater construction complexity.

In addition to the downstream constant level gates like those mentioned, there are others operating in the same fashion and which are used for bottom turnouts in reservoirs or canals located at a higher level. This is shown in Figure 19.14.

These gates require stilling basins that will dissipate the energy of the water passing through the gate. A practical rule is for the stilling basin volume to be equal to $21 \cdot Q \cdot H^{3/2}$ where Q is the maximum rate of flow and H is the water drop.

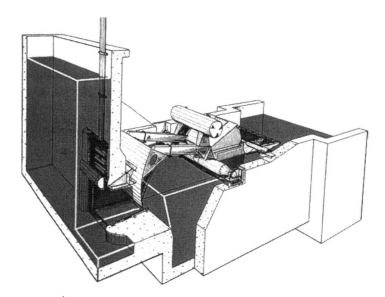

Figure 19.14 Bottom turnout with a downstream constant level gate.

It is recommended that the stilling basin length be at least three times its width and the width be one and a half times the water height. If the gate dimensions impose a different stilling basin size, then this should be corrected as necessary, but without any reduction in volume.

19.7.3 Automatic mixed constant level gates, upstream and downstream

Fixed level gates located downstream that we have just described are, undoubtedly, an excellent solution from a theoretical point of view. There are, however, situations in feeding the canal, in which these and other gates, involve conflicting problems, such as the case in which insufficient flow rate reaches the gate so that, in order to provide the corresponding pre-established flow downstream, the upstream section of the canal would be emptied, an event that is highly undesired. In a similar fashion, if too much water arrives and the gate continues to attempt to maintain the pre-established value downstream, the canal overflowing would occur upstream.

Mixed gates operate in the following manner:

There is a minimum water level upstream from them and so they completely close to prevent the canal from emptying. When, with the minimum level upstream, the flow rate begins to increase, and until it reaches the required level downstream there is a transition operation during which the difference in level between upstream and downstream is a pre-established one.

From this upstream level, and even if this rises, the gate will maintain a constant level downstream equal to that required. This situation will be maintained until a maximum value upstream, or higher values, apart from which the gate will fully open in order to prevent any overflowing upstream.

The design of these gates is both ingenious and complex. They consist of a circular board, connected to identical circular floats that are rigidly welded to the gate shaft and which are partially submerged in two basins of water, one upstream and the other downstream from the gate (Figure 19.15). The arrangement is such that when the difference in level in the two basins is equal, the resulting torque is fixed, independently of whether both levels rise or fall by the same amount.

The previously described operating curve is achieved because there is a series of connecting spillways between the basins and the upstream and downstream levels, in addition to other connections between the basins which, on working in the various possible situations, cause the gate assembly to operate in the desired fashion.

The Figure 19.16 is a photograph of this type of gate in the Toro–Zamora Canal, shown at the time of its installation.

In the same way as fixed level gates upstream or downstream, mixed gates produce maximum benefits, not when they are installed singly, but instead when several are installed in the same canal, which will allow excess flow rates released by one gate to be stored, up to a certain limit, by those further downstream. Or when a gate is closed due to a lack of feed, those

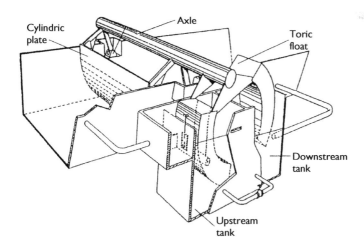

Figure 19.15 Sketch of a mixed gate (Einar).

Figure 19.16 Mixed gate in the Toro–Zamora Canal (Einar).

downstream will be able to handle the users' needs by supplying the water held in storage.

Of course, the choice of gates, their locations and separation, etc., would not be an easy problem to optimize if it were not possible to employ mathematical simulation models, which we shall look at in Chapter 20.

Automatic mixed level regulation gates have been employed much less than those with only a fixed level upstream or downstream. The electrical gates that we will examine a little later, can produce the same objectives and even more. The limiting factor is that these require the laying of an electric power line or other electric devices, which can be costly in isolated areas and also quite vulnerable to storms.

19.8   Other considerations with respect to automatic hydraulically controlled gates

These gates will maintain a constant level either upstream or downstream as applicable, based on constant readjustment in accordance with the oscillations in level between which some perturbations are found, such as waves produced by the wind.

For the purpose of reducing the instability that this effect could produce in one of the constant level gates upstream, which are very sensitive, a shock absorbing system, something similar to those vehicles have in their suspension, is installed. The gate movement is then braked and stabilized.

However, throughout several years of installing and operating automatic gates, we have had certain problems with shock absorbers, because they were prone to heavy corrosion. The wet environment in which they are installed, together with the differences in ferric material they and the gates are made of, is undoubtedly the cause of galvanic corrosion, which requires strong protection for these elements, which are, on the other hand, essential. In fact, we remember a case, in which the operation of a derived turnout distributor caused the fast response of its associated regulation gate. This created a wave, either rising or falling, which, apart from being transmitted along the canal, modified the apparent water height close to the gate, which, in turn, produced a new response, which was summed with the previous one leading to instability. The installation of a suitable, more rigid shock absorber resolved the problem.

The possible effects of resonance can also be resolved in this way. This sometimes occurs when a wave produced by the gate propagates along the canal and is reflected by some obstacle, for example a syphon inlet or outlet installed in the canal. The reflected wave could reach the gate at a time that tends to increase the remnant oscillation. The variation in shock absorber hardness will vary the gate period and prevent the danger of resonance.

The constant level gates downstream can employ a more efficient system. The float moves with the gate and is housed in a metal box, which is closed at the top and fixed to the canal. Air is able to enter and escape through a narrow slot. The significant head loss through this means that the gate movements are slow.

19.9   Duckbill weirs

Constant level gates are more vulnerable than a civil works construction, especially in the smaller sizes. For this reason, in the Orellana Irrigatable Zone, where there is much development in the use of automatic hydraulically controlled equipment for canals, very strong elements have been designed to replace constant level gates upstream.

These consist of very long, fixed crest weirs that are installed in the canal. Since the weir responds to the following formula:

$$Q = C \cdot L \cdot h^{3/2}$$

where Q is the circulating flow rate, C is a constant close to 2.1 for the Creager profile, L is the crest length and h is the spilled water thickness then h can be made small enough if L is sufficiently large.

For flow rates that are less than the maximum, the spilled thickness will be less than that chosen h, and the water level upstream will oscillate between the crest level and this same height increased by the value of h; in other words, it will vary only a little, so that the turnouts upstream will be regulated.

In order to obtain a sufficiently long weir length L, this is constructed diagonally to the canal, and to reduce the affected zone it is given a symmetric form, so that it appears somewhat similar to a duck's bill, to which it owes its name (Figure 19.17). Figure 19.18 shows a civil works drawing taken from a standardized collection of works (72).

In order for it to work correctly it requires a certain amount of care and attention from the hydraulic point of view.

Figure 19.17 Duckbill in Orellana Canal.

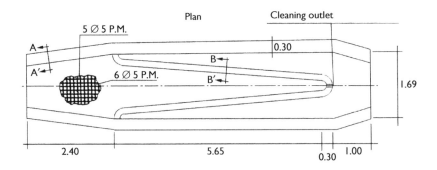

Figure 19.18 Duckbill plan (Liria and Torres Padilla Collection).

It should not be forgotten that the water being transported by the canal does so at a speed parallel to its axis and therefore obliquely to the weir crest. However, the formula given for the weir is only valid if the lines of current flow, when they drop, are normal to the crest, which is the same as saying that they are not influenced very much by the approach speed. If this is not the case, the benefits of having a long weir length are only apparent because liquid flow over a differential element of area only takes into account the speed component normal to the surface it passes over. This is due to the fact that that speed component parallel to the surface it passes over only transports the moving particle over this surface, but does not make it pass through.

The best way of reducing the approach speed is to increase the section.

For this reason, the canal is usually widened at the duckbill weir so that the speed is reduced to a maximum of 0.50 m/s, and even less on occasions.

Duckbilled weirs are used to operationally replace gates that provide constant level regulation upstream. While they can guarantee a practically exact level, they do have the disadvantage of only being able to ensure this with an oscillation h, so that the level of oscillation at a derived turnout will be increased by this amount, with the possibility of the distributors not being able to absorb it. This can result in the stretch regulated by each duckbilled weir being shorter than when gates are employed.

On the other hand, it has to be said in their favour that they are robust structures and can be constructed by personnel with little specialization. Our opinion is that they can be recommended for relatively small canals, because the equivalent gate is quite delicate due to its small mass. On the other hand, gates in large canals are heavy and can withstand rough treatment by the users in a situation in which duckbilled weirs are too large and less recommendable.

In order to partly resolve this problem within certain limits, a bottom gate may be combined with a duckbilled weir, to permit water to pass downstream along the canal.

The theory of operation is as follows: the flow rate in the canal varies in two basic ways. One is the variation in demand over a period of one day. The other is the seasonal variation according to crop development. The first can be absorbed by the variation in water thickness spilling over the duckbilled weir and the second is regulated by further opening or closing the discharge gate, which is only necessary with relatively long time intervals.

In this way, with a relatively short duckbilled weir, the oscillation can be reduced to small values, although it has to be remembered, as already explained, that the approach speed, even with maximum flow rates, must not reach 0.50 m/s, which limits this method in practice.

These duckbilled weirs should not be confused with certain transverse weirs that are sometimes installed close to regulation gates. Figure 19.19 shows the Guarico Canal in Venezuela, with two spillways, one on each

Figure 19.19 Safety device in Guarico Canal (Venezuela).

side of the regulation gate. If some slightly incorrect operation is performed, or oscillation occurs in the canal, etc., then the level upstream will tend to increase, producing a spilling of waters towards the canal downstream. It is in fact a supplementary safety device against overflowing, but it simply does not have sufficient crest length to maintain a fixed level that would regulate the canal.

19.10 Automatic electric gates in response to annexe variations

The simplest of these can be found in the Western United States, where it caused great commotion in the world of irrigation. It is called "Little Man" and it was intended to search for objectives similar to those of the previously described hydraulically controlled gates, but with more flexible operating possibilities.

In its simplest form, Little Man was intended to achieve level variations close to the gate that lay within a pre-established range. It made use of a float fitted with electric contacts on pulleys for operating a motor that would drive a gate open or closed. This movement would be carried out in short pulses, controlled by a timer switch in an attempt to prevent instability.

Since it is very easy to change the length of pulley transmission, the reference levels were also very easy to modify.

The possibilities of this type of gate were enormous, above all, by perfecting two components: the sensors employed to measure the basic gate control variables, which were generally levels, but could also be flow rates and degree of gate opening, and other electric components that would automatically react to these measurement variations and transmit operating commands to the gates in accordance with pre-established operating criteria. These are the so-called "Programmable Logic Controllers" (PLCs), some of which are described below.

19.11 Data sensors

The basic variables in the operation of a variable regime canal are the levels and flow rates or speeds. When gates are involved, it is also necessary to know just how much they are open or closed.

These measurements are often interrelated so that if some of them are known, the others can be calculated quite simply, above all, when performed on a computer or programmable calculator.

For example, if the level and section shape are known the hydraulic pass area can be calculated, which makes it irrelevant whether the flow rate or speed is known because the other measurement can be very quickly established.

If the water level passing over a thin or thick wall measuring weir is known, then the circulating flow rate can be immediately calculated with a suitable formula.

If the degree of opening of a gate is known, together with its shape and dimensions, then the pass area can be exactly established, and this, together with the levels upstream and downstream, can be used to calculate the circulating flow rate, without needing any specific measurements.

This shows that, in general, there are two commonly used types of sensors, those to measure levels and those to measure how much a gate is open.

The simplest form of level sensor is the previously described float. The pulley causes an analogue signal to vary by, for example, acting on a resistor connected to a dc circuit, the current measurement of which is a function of the water level in the canal.

Another way to measure a canal's water height is by means of a metal manometer which measures the water pressure above it, producing an analogue signal. The manometer accuracy can be improved by combining it with a precision balance.

Another successfully employed system is that of permanently injecting air bubbles at the bottom of the canal. The air pressure inside the pipe is equal to the water column acting on the end of the pipe, in other words it is equal to the water height. The air flow rate is so small that it does not establish any difference in pressure due to loss of head. The pressure is measured at an intermediate point along the pipe, between the air entrance and the exit.

This procedure presents a series of interesting characteristics. It is very important to point out that it completely lacks any moving parts in contact with the water, such as floats, cables, etc. It is a system that is not affected by the presence of ice or any other exterior circumstances. It is highly advantageous for the air circulation system to be open and so leak-tightness requirements in the pipeline are not excessive. One inconvenience is the requirement for a small automatic air compressor fitted with a small regulation reservoir. This may be replaced by a tank of compressed air which obviously will have to be periodically replaced.

Level sensors that are based on employing the different electrical resistance between two submerged electrodes, depending on the water height, normally involve still unresolved problems of algae growth, particles in suspension and dissolved salts, all of which have a very negative effect on measurement accuracy and guarantees.

Sensors designed to measure gate aperture are based on the fact that it is always possible to associate this measurement with an angle of rotation, which is quite evident in valves and gates operated by wheels and in segment or circular sector gates, and it is also true for flat board gates with vertical movement and driven by chains because these operate with pulley or toothed wheels, the rotation of which is proportional to the movement.

The employed measurement equipment consists basically of a converter that produces a dc voltage signal, for example, 4.20 mA, which is proportional to the angle of rotation and which is subsequently converted into a digital signal.

With regard to flow rate measurement sensors, we can state that not all are based on the previously described possibility of obtaining them with respect to level readings, some are capable of providing them directly.

The most typical of these is the rotating propeller measuring structure. This consists of a wheel with blades that rotate under the effect of the water flow, at a speed that is proportional to the water's own speed. A gear mechanism drives a speed indicator and simultaneously produces an analogue signal which, as always, may be converted into a digital version, if so required.

The problem of this type of equipment is that it measures the speed of the water with which it is in contact and, as is now well known, the various points throughout a canal's transverse section have different speeds. For this reason, they are usually employed in any available inverted syphon pipes, in which the speed is much more uniform (Section 15.5).

Nowadays, devices are also employed to measure flow rates based on ultrasound, which were described in Section 15.6.

Sensors can be used to measure all the variables that are required to control gate movement in one way or another. Another component is needed that can react to the signals transmitted by the sensors and, in accordance with pre-established rules, is able to drive the gates in the form that we desire. This component is the Programmable Logic Controller, or PLC, which has transformed industrial process engineering and is of great use in irrigation systems.

19.12 Programmable Logic Controllers (PLCs)

Because of their incredible possibilities of usage, PLCs are truly outstanding in the regulation of canals among those components that are physically capable of providing automatic regulation.

In the same way as digital computers, these components are based on microprocessor circuits that employ Boolean Algebra. They can perform practically any industrial process control task.

They have a series of digital or analogue inputs, together with a series of outputs. Their outputs can be programmed in function of their inputs, and these outputs are then used to provide commands to the industrial process equipment.

They usually make use of PROMs (Programmable Read Only Memories), in which the application control program is stored. This does not normally undergo any modifications, or very few, during operation. The circuits also include RAM (Random Access Memory), which is used to temporarily hold status variables, timings, etc.

The system may also be expanded by peripheral data outputs, such as monitors, disk drives, printers, etc.

The characteristics that best define the excellent performance of PLCs are as follows:

- Reliability
- Flexibility
- Ease of operation and programming
- Power
- Economy.

Until recently, the main difference between PLCs and digital computers was that the former basically performed logic operations, while the latter could also carry out arithmetic operations, although there are now PLCs available that can perform arithmetic operations too.

Moreover, in the case of digital computers, several languages are available for programming purposes, such as Fortran, Algol, Basic, Cobol, PL1, C++ and Pascal (Delphi), etc. However, for PLC programming it is necessary to follow the individual manufacturer's method, which almost always falls into one of the following types:

- *List of instructions.* This is a set of symbolic codes, each of which corresponds to a machine code instruction.
- *Contact diagram.* This system is used to establish the tasks to be carried out by the PLC using an equivalent electrical contacts diagram.
- *Function diagram.* This is a symbolic language in which the various combinations of variables are represented by standard logic symbols.

- *Graph system.* This system makes use of the property whereby any flow diagram for an operation corresponds to a PLC machine code program. In accordance with this, some manufacturers have designed applications to translate these graphs into a PLC program.

A computer with the corresponding application can be extremely useful in the programming of a PLC, which has the additional advantages of being able to store the program on a disk and list the program on a printer, together with embedded remarks that assist in its understanding.

The usefulness of PLCs as regulation components in canals is quite obvious. Information from the sensors that measure water levels, flow rates, gate apertures, etc., is fed to the PLC inputs. Commands are transmitted from the outputs for the control of the gate motors, either opening or closing them, usually by means of pulses, in other words, by successive discrete commands that permit slowly approaching the final position, without causing further problems due to too fast an operation.

The input data and output commands may be either analogue or digital signals.

Because, in general, the PLC is prepared for analogue signal inputs and outputs, this would appear to be the simplest system. However, these can only be employed over short distances between the sensor and the PLC, or between the latter and the motors. This is because the analogue signal is attenuated by the conducting wire resistance.

In the second case, modulators–demodulators (modems) are required, which convert analogue signals into digital ones or vice versa. These have a great advantage, in that digital signals are not affected by distance.

The main application of PLCs is providing automatic gate operation with respect to local variations. A simple example is that of a gate intended to maintain a constant level. This requires a sensor to provide a signal indicating the actual water level and, assuming that this level is incorrect, the PLC will respond to this by issuing the corresponding command for the gate to open or close as applicable.

The advantage of PLCs is that they can accept programs that are much more complicated, which means they can control much more perfect and complex operational programs than those for simple hydro-mechanical regulation gates. On the other hand, they are much more prone to the effects of lightning, data transmission problems, etc.

19.13   Possible automation methods

PLCs have three interesting ways of programming the output commands and these are basically known as "proportional", "integral" and "differential".

Using the proportional system, the gate commands can be made proportional to the measurements provided by the sensor or to others derived from them. For example, the gate aperture can be made proportional to the water level, flow rate or speed.

The integral type of programming can make gate operation depend on the sum of data for a specific measurement, for example the flow rate, over time. In this way, it becomes possible to control the gate which tends to correct the integral of the errors between the flow rates that should be circulating and those that are really passing through the gate.

Less used, but still interesting, is the differential command, which establishes a reaction at the output as a function of the input signal variation. This allows a gate aperture to be established that is proportional to, for example, a drop in level, therefore responding to urgent situations.

There are many publications available that provide further information on this very interesting subject (47).

The ability to transmit information over long distances by means of digital signals and modems makes it possible to operate a canal in an almost perfect fashion, operating the gates in order to resolve problems that are known to be about to occur, because it is known when they commence at remote locations. Any perturbation at a given remote point will produce a surface wave in the canal that is transmitted at a specific velocity along the canal, until it reaches the gate that is to react in order to correct the original perturbation by either opening or closing. If the gate were to react because of signals coming from sensors located close to it, then it would not actually react until the effects of the perturbation had arrived, which would require a more or less long time to actually arrive. If, on the other hand, data transmission is carried out using digital signals, the gate would be able to react instantaneously.

Gate reaction can be performed with respect to changes or stresses occurring either upstream or downstream, known as respectively as "upstream control" or "downstream control".

However, the idea of making the gates work in combination is very useful, for example arranging for the water heights of the two ends of a stretch comply with conditions of the type $\alpha \cdot Y1 + \beta \cdot Y2 = K$, where α and β, together with K, are coefficients and $Y1$ and $Y2$ are the water heights. It is easy to appreciate here that if the canal is rectangular, α and β are equal, the condition is that the water volume in the section is constant. If they are not equal, it is also equivalent to a certain condition with respect to the volume, adapting very well to the concept we described at the beginning of this book of a canal being a mixed transport and storage structure.

A thorough examination of modern canal operational possibilities is provided in (26) together with a magnificent publication on hydrodynamic regulation methods for canals in (81).

If a very large number of interconnections are to be made between the gates in order to establish overall canal operation rules, regulation should be performed by a computer. This is capable of receiving and processing

Figure 19.20 Electric control gates in the Genil-Cabra Canal (Spain).

much more complex rules through adequate applications programs, and will then be the overall controlling element for the installation. Figure 19.20 shows some electric control gates in the Genil-Cabra Canal.

The automatic operation of a canal, even more so in the case of a canal network, requires a mathematical model that enables the hydraulic operation to be simulated, in order to permit prior correction of the automatic commands we are to implement. This point, which is extremely important for the new conception of canal operation, has been the subject of millions of hours of research by many specialists, and we shall speak more of it in Chapter 20.

19.14 Reservoirs

A very useful element for canal regulation is the reservoir connected to it. These are used to store more or less quantities of water when there is surplus water and to return it to the canal when required. It has the disadvantage of requiring topographically suitable locations for their construction, otherwise they would be prohibitively expensive; however, when these locations are available, one can be quite sure they will be beneficial. They can be designed to be stand-alone devices or in combination with automatic gates, which will noticeably increase their performance.

Since time immemorial, reservoirs have been constructed close to irrigation canals for the purpose of storing water during periods the canal is delivering to the smallholdings, in order for this to be used later when necessary. Neither conception nor design present any great difficulties and their aim is not so much to regulate the canal, but to regulate the water delivered to the irrigators.

We are not referring to this type of reservoir, but instead those which have a direct effect on the canal, from which they receive water and to which they return it. They may be interleaved with the canal or located to one side and connected to it by means of a turnout operating in both directions. This single connection is sufficient because the reservoir water surface slope is practically zero since the canal hydraulic gradient is very small, but it will be even less in the reservoir because it is much wider in order to attain the desired volume.

In any case, it is important to have a drop of difference in level located in the canal. For optimum reservoir design, its water height must coincide with the difference in level existing between the level upstream and the canal level downstream. Under these circumstances, the effective volume is at a maximum, with an available surface areas determined for the reservoir.

These reservoirs can have significant capacities; the two designed to regulate the dead section on the Zújar Canal and invert the turbine timetable for the plant located in the canal feed have capacities of around $300\,000\,m^3$. Similar capacities are planned for the Los Payuelos Canal and $50\,000\,m^3$ for one on the Inés Canal in Soria, and also the irrigation canal tail on the Zújar Canal.

The necessary capacities for reservoirs can be very large if they are intended to invert or greatly modify the circulating flow rate law. However, their purpose is often only to regulate canal operation which, due to operating errors, unexpected or other circumstances, undergoes variations in its circulating flow rates. In many of these cases, the construction of discrete volume reservoirs can be very effective.

Reservoirs can be constructed by taking advantage of a natural watercourse, which is closed off by a small dam or mainly excavated and part of the perimeter closed off with a dyke constructed from the removed soil, which is the solution for flat areas.

Canal impermeability is extremely important and a sufficient level of impermeability is also necessary for a reservoir. If the underlying ground does not have this necessary level of impermeability, then it will have to be lined. The same type of lining as for the canal may be employed for the reservoir, but this is not essential. In fact, the use of membranes, plastics or asphalt is a widely used, economic solution.

We believe that the most effective solution is a reservoir that is fed from the high part of the canal drop, combined with a flow rate regulation system at its outlet at the lower part of the drop. An automatic, constant level gate downstream, with adjustable flow rate distributors located in the gate stilling basin, is highly efficient. The reservoir is thus fed using the arriving surplus flow rates. The canal feed flow rates at the outlet are fully regulated by the distributors. This solution was designed for the Inés Canal.

If the drop is small, the effective water height will also be small; the same applies to the effective volume, meaning efficiency is also small.

While it is true that if there is a difference in level available, then it would be possible to install a pumping system; however, this is complicated and increases installation costs and hence this is not frequently used. This solution may be adopted when there is a clear site possible for the reservoir which is located at either a higher or a lower level than the canal.

Reservoirs that are constructed interleaved with the canal will intercept all suspended solids, whereas those constructed to one side will be almost free from them. In the former case, some form of cleaning system is required, which would ideally consist of a deep discharge outlet leading to a nearby river or ravine. A sudden high flow rate could then be sent, which would drag along with it all the accumulated sediment in a fashion similar to that described in Section 11.8. Because of the fact that in many cases the accumulation of suspended material at certain points can produce serious problems, it might be a good idea, if there is a suitable cleaning discharge outlet, to use interconnected reservoirs with the double function of canal regulation and sedimentation operations.

The employment of mathematical simulation models, which are described in Chapter 20, will greatly facilitate the suitable design of these structures, taking into account the storage volume, the degree of regulation obtained, etc., which would otherwise be practically impossible to produce.

A very interesting reservoir case are the so-called "pressure chambers". These are constructed at the end of canals for electric power generation, from which forced piping is taken to the hydroelectric plant's piping. Their purpose is to regulate the arriving flow rates which, because of the canal's length, are modified with often significant delays in relation to the supply request command issued by the plant and the aperture variation of the canal's feed gates.

Although the use of flowing water plants is decreasing in Spain, they continue to be frequently employed in developing countries.

Their design and construction do not involve special problems with respect to other reservoirs; however, it is very important to give thought to their operation.

When the pressure chamber is being designed, the most important aspect is to calculate the necessary volume. Figure 19.21 shows the plan layout for one. If the chamber is already in operation, it is very important to calculate the drop's possibilities under several operating hypotheses in order to optimize its operation.

Both these studies can be very well carried out using the mathematical models described in Chapter 20, which contains a very interesting regulation example for the Zújar Canal employing reservoirs.

Figure 19.21 Pressure chamber for a hydroelectric plant.

19.15 Centralized control canal regulation

Modern canal regulation includes two basic operations: (a) information on gates opening, the level and/or flow rates at important points and (b) the taking of decisions on operations in accordance with existing problems and the irrigation service that is to be provided.

Both operations are very important and it is desirable that both are centralized at the same location, which is the local centre for the canal network operations control, with the supervision of the Operations Manager.

Information on the current situation is continuously provided by the sensors that are installed at important locations along the irrigation network, which read and transmit level and/or flow rate data by means of analogue signals. This information is transmitted over cables or via radio, depending on the actual case. Since the analogue signals can only be transmitted over distances that are invariably too short, a modem is required to transform them into digital signals and these are the ones that arrive at the control centre. This will require an electric power supply, which is normally obtained from batteries or solar power generators.

The information received at the control centre is computer-processed, which then displays a scale graphic of the irrigation network containing the level and flow rate data, together with storage volumes, if this is required and also providing data outputs, for example, to printers, plotters and, above all, disc files that store all this information.

This centralized information can be employed with hydro-mechanic as well as electrically controlled gates governed by sensors, whether the readings are measured at the gates or at some points remote from them. They are of great use to the situation databases for the canals that produced them as a source of data for future operational plans. They are also extremely useful in the rapid detection of abnormal canal conditions that could be indications of significant problems. Such is the case of rises in water levels, which could indicate some form of regulation gate obstruction or faulty sensor operation. They may also indicate large drops in water levels, possibly indicating the collapse of a backfill.

If the centralized information installation is exactly as indicated, then the canal network Operations Manager can only take action by sending a 4×4 vehicle with a team of engineers to the location to carry out any necessary repair work, or can telephone any employee who is close by to perform an inspection so that rapid and precise information can be obtained with respect to the true situation.

However, there is also the possibility for the gates to operate, not automatically in response to hydro-mechanical effect or sensor data, but, instead, for all or some of these gates to receive commands directly from the centralized control centre. This will require certain control circuits, similar to the previously described ones that will transmit the Manager's commands to the gates. The command will be transmitted in digital format, and a modem at the gate will convert this into an analogue signal. A PLC then uses this signal to produce an electric current that will drive an electric motor to open or close the gate, generally in progressive increments. This also requires electric power that is normally supplied by batteries or solar power installations if there are no suitable mains power lines available. If there is no other power, then the batteries will be charged by small motor-driven generators that are periodically started up.

The difficulty associated with this situation is for the Canal Operations Manger to decide on what commands are to be issued. For this reason, the use of a computer mathematical model of the type described in Chapter 20 is essential. This, as a function of the current canal conditions and the possible commands that can be issued to the gates, will provide a real time response of the resulting levels and flow rates in order to establish whether a given command is suitable or not. This can also be carried out in a simpler fashion by having the computer perform the necessary calculations to be carried out by the PLCs described in the previous solution, so that set points for each gate are achieved.

In such an installation, there is a dialogue between certain points along the canal and the centralized control centre. The former will transmit information to the centre, which will respond by sending a series of commands that will produce the necessary gate movement.

The Provenza Canal in France has been operating in this fashion for over thirty years, with a mathematical model that is one of the first to be installed, at least from among the ones that we are aware of.

Chapter 20

Mathematical models in canals

20.1 Introduction

Canals function in a variable manner; the situation of their levels along their lengths and the changing flow rates they transport at any given moment at the various points are not at all easy to predict; and even less is known about how they are going to react with respect to variations in the canal elements, for example, due to the modification of gate apertures or type of feedwater flow rates or those demanded by secondary canals.

It is a good idea, we would say necessary, to have a mathematical model available, which will permit each situation to be simulated and at any given time, in order to establish how the canal is going to react to these modifications. If an already constructed canal is now in the operational phase, such a model would be of great assistance to the canal operations team by helping them to establish whether a specific operation that is to be carried out is going to achieve its objective without any problems or, on the other hand, is it going to lead to overflowing or a lack of water at certain points along the canal. This requires a mathematical model with a fast response in order to provide this information in real time and, where necessary, allow the operations manager to put into place any other measures that are considered necessary.

However, if the canal is still in the design phase, a provisional mathematical model can be used to establish the advantages or otherwise of two or more different solutions that may be adopted for the canal gates or other installations, its level of use can then be enormous.

The amount of understanding gained by those responsible for the operation of the canal should not be underestimated either. Putting the model to work under varying situations can provide a great deal of information about its operation.

The subject of mathematical models for canals is highly specialized and complex. It has lead to huge numbers of books being produced and has occupied the time and energy of experts in theoretical hydraulics, engineering, mathematics and software programming (10), (11), (12), (40), (61), (62), (63), (75), (79), (88).

We cannot venture as deeply in this subject, in a book intended to provide a general overview of canals, as many readers might feel is necessary. But, since engineers working on canals, in all their variations of design, construction and operation, will very probably become involved with some form of mathematical model, we feel it is essential to provide at least some general information.

20.2   The bases for mathematical canal models

A mathematical canal model should commence with the mathematical expressions that govern water movement. For each infinitesimal element of canal length there are two conditions to be met: one referring to the continuity of the water mass and the other to compliance of the Theory of Amount of Movement, which relates forces and speeds. In this case, the general theorem should be applied because the amount of movement is modified, not only by the variation in speed of the infinitesimal element, but also because its volume varies in function of the water height because of its variable regime.

Both the conditions are expressed as two differential equations. The system of these differential equation can be subjected to current mathematical rules of sums and combinations producing two equations classically known as those of Saint-Venant, which were mentioned in Chapter 1.

These equations relate water movement properties at infinitely close points. It is not possible in engineering practice to deal with infinitely close points, their differentials must be replaced by finite increments that are sufficiently small so that the results of the calculations coincide in practice with the true situation.

For model purposes, the canal is divided into small-length sections, which depend on the canal type, its total length, gate numbers and form, together with any special works constructed in the canal itself, etc. and, in order to define ideas for the reader, we will say that they are only a few hundred metres in length.

The model is intended to establish flow rate, or speed, and water height variations over time in infinite increments, at all canal points during short intervals. We speak of flow rates or speeds because there are models that deal with speed and others with flow rates, which does not really matter, because by knowing one of the two, the program can calculate the other based on the existing water height.

If we commence with an initial situation in which we are aware of something like a photograph of the canal, in which all the levels are marked and of another representing all the flow rates at each point, the model must be able to provide the information corresponding to successive photos over various time intervals.

For this reason, it begins with the fact that the Saint-Venant equations will relate water heights and flow rates at contiguous points with the same variables as in the following time interval by replacing the derivatives with finite differential quotients.

20.3 Calculation structures for mathematical canal models

The Saint-Venant equations form a system of second degree partial derivative equations, which requires complicated mathematical processing and which has to be performed for all the end points for the sections into which we have divided the canal for calculations purposes.

These calculations must also be repeated for each time interval in which the new flow rate and level situation is to be established, which must also be small intervals, so that there is only a very small error in relation to the true situation. The information that the model can provide is that corresponding to all the selected points in all the calculated time intervals. However, such a volume of data would be impossible to handle in practice, apart from the fact that the difference between two close time intervals would not have a great deal of interest. For this reason, although the model actually calculates the phenomenon for very short time intervals of, for example, only a few minutes, in order to attain sufficient precision, it usually only presents the results for periods that are multiples of this time, for example each hour.

It is quite evident that such a large volume of very complicated calculations could only be performed on a computer, employing very well designed complex programs. In order to produce such programs, the mathematical methods to be employed must be carefully examined.

There are two basic mathematical methods for the problem processing from the computer point of view. One of these is the Characteristics method and the other is the System of Implicit Equations.

The characteristics method, which has been widely covered by Wylie (12), (88), replaces the Saint-Venant differential equations by another two for which the graphical representations commence with two contiguous points in the canal, with levels and speeds or flow rates that are known for the previous interval. The corresponding intersection will give us the flow rate or speed values and water height at a later interval at an intermediate point. It is generally necessary to use interpolation in order to obtain data for the exact point under study.

The program based on the Characteristics method is relatively simple and only requires computers with modest memory specifications.

The results are not always satisfactory for canals because, apart from the problem of the necessary interpolation, the system may become unstable (61), with enormous, completely illogical water level heights appearing at certain points. It was Courant who established the mathematical condition to prevent this and which lead to a relationship between the time interval under consideration and the space between two consecutive points, which forced a reduction of the calculation interval and leading to longer total calculation times. The Courant condition is that the chosen calculated time interval at all points must be less than the time taken by a perturbation at one point to reach the next point, and this involves the employment of short calculation periods.

The other system of tackling the mathematical problem of obtaining the successive photos which, over time, define the water heights and flow rates at the various points throughout the canal is defended by Preismann by means of resolving a system of implicit equations.

This method is based on the resolving of a series of equations for obtaining each of what we are calling photos of the canal.

The system of equations is obtained by applying the conditions of Saint-Venant to all the intermediate points chosen within the canal, expressing the partial derivatives as a function of not only the known values for the preceding interval, but also the unknown values for the following interval. The values of each point are related to those of the two contiguous points, which requires the resolving of the equation system. The most suitable applicable method for resolving the system is that of Newton-Raphson.

From the numerical calculation point of view, this method is more stable and does not usually require the condition of Courant and it has given us superior results. There are, however, many models based on the characteristics method that work perfectly and it is a method that does not have as severe memory requirements as the implicit equation method, because it implements and resolves the Saint-Venant equations for each internal section of the canal, between each two consecutive points, one after the other. We have found this to be the slowest method, but with modern computers, neither the necessary amount of memory nor the calculation speed poses any problems.

Each gate within the canal represents a point of hydraulic discontinuity, where the Saint-Venant equations are no longer valid and where, in order to replace them, equations defining the gate type have to be established. Something similar occurs with special works installed in the canal, such as inverted syphons, etc. Summing up, each installed gate or other works is, from the engineering point of view, an element or structure possessing specific physical and hydraulic characteristics, but, from a computer calculation point of view, is represented by two equations that define it and replace those of Saint-Venant.

For example, a constant water level gate upstream, located at point i in the canal, which fixes a water height $h(i)$ with a flow rate $Q(i)$ and which also has a lateral turnout with flow rate $q(i)$, is represented by two equations as follows:

$$h(i) = K(\text{constant})$$

$$Q(i+1) = Q(i) - q(i)$$

This will require very careful study by the programmer of the equations that are going to represent the gates and other concrete constructions installed in the canal.

Once the equation system, consisting of the Saint-Venant equations referred to the normal canal points and the equations for the gates and other constructions in the canal referring to points with specific works, has been prepared, it is then necessary to check that the planned resolution system accepts the form of the equations at points with gates or other constructions because, since they are very different from those of Saint-Venant, the calculation system might abort due to a division by zero.

One frequent problem with mathematical models for canals is the presence of discontinuities which, as previously stated, can lead to ridiculously high water heights, causing the program to abort (70).

The cause of this can be found in the fact that with the imposed gate operating conditions it is possible for transitions to occur from slow to fast regime or vice versa, with the appearance of a hydraulic jump. The Saint-Venant conditions are not met at this point from a mathematical point of view and so the situation must be taken into account. The way to do this is when the Saint-Venant equations are transformed into discrete versions in order to form the implicit system, they are not used directly, but, instead, the integral form is called, in other words, the equivalent form of the equations that would result had the differential equations been integrated. The reason for this is that the Saint-Venant equations handle partial derivatives, which are functions that are not continuous in certain hydraulic phenomena, for example jumps, on the other hand, integral equations are continuous.

20.4 Boundary conditions

The canal model must refer to a specific canal and requires the introduction of data for shape and dimensions of the transverse sections, rugosity factors, gradients, types and dimensions of special constructions installed in the canal, such as inverted syphons, localized head losses, lateral turnouts, jumps and measurement structures, etc., together with the gates and their types, whether constant level, upstream or downstream, controlled by flow rate or levels from a sensor located at a different point, operating with predefined temporary laws of levels of flow rates, etc.

There are, however, still further data to be introduced into the model, which are usually called "boundary conditions".

If, for calculation purposes, we divide the canal into n stretches or sections, this means that including the ends there are $n+1$ points. Since there are two variables for each point, which are the water height and the flow rate or speed, there will be $2(n+1)$ variables. But for each section that links two internal points there are two equations, so there is a total of $2n$ equations. There are, therefore, two variables plus the $2n$ number of equations, which implies that in order to be a determined system, two variables have to be arbitrarily fixed, which are, in fact, time functions.

In normal canals operating in subcritical regime, one of the variables is the entry flow rate at the beginning of the canal, which is a data item, and another is the flow rate or water height at the exit. The water height could also be given for the entrance instead of the flow rate, although this is less frequent. With a canal in subcritical regime, data values have to be provided for the beginning and the end of the canal because with subcritical speed, the water movement is influenced by what happens both upstream and downstream.

This does not happen with a canal in supercritical regime, because at supercritical speed, water movement only depends on what happens upstream and whatever happens downstream does not affect any of the points. The functions to be defined therefore are the flow rate and water height at the entrance at the beginning of the canal.

These facts are of great importance with respect to the method of resolving the implicit system of equations. With a subcritical regime, a double sweep must be carried out, which means that, first of all, each equation is combined with the next until the last is reached, which contains data we have fixed ourselves. The second sweep is the second pass, from bottom to top, which produces the water heights and flow rates for all the points in function of the data element we entered at the final point.

On the other hand, in supercritical regime, the two data elements we fix are located at the beginning of the system, which means a single sweep is sufficient.

The system resolution methods are not exactly the same for both the slow and the fast regimes, which introduces a complication into the calculation for a canal with sections of different regime. This is undoubtedly the reason why many mathematical model computer applications for canals do not include this possibility.

20.5   Model characteristics that define their quality

The model must run quickly on the computer so that it can predict the canal's response to any operation in real time.

The results it provides must have sufficient accuracy and they are sometimes subjected to an easy-to-execute test. A canal is designed with a horizontal bottom and horizontal water surface level with zero flow rate. A flow rate with a certain temporary law of variation is introduced at the main turnout and, at the same time, it is emptied at the end with an equal law. A sinusoidal flow rate law is frequently employed, including the full wave, so that the entry volume is the same at the exit. The final volume should be equal to that at the beginning, with any difference being an indication of program accuracy and this is normally less than 1% of the entry volume.

Stability is a very important condition and involves, among other, two different conditions. The first is that the program should not abort, which means it should not hang up computer operations if an absurd situation is reached, for example, if a canal section becomes dry. The second is that the results should be smooth, without any sawtooth variations that do not exist in reality and, when different data are introduced, the results gradually vary. A lack of stability usually indicates a bug in the model's program.

Another condition that any good model should have is that it must be easy to introduce the initial hydraulic condition of water surface and flow rate at each point. A start is often made with a simple situation of water surface parallel to the bottom or zero flow rate with horizontal level. The flow conditions are varied so that at the end of a first test, the situation is similar to the initial one we require. The model must be able to save in a file the flow rates and water heights for each point and also be able to restart the calculation from this initial situation. It must not be forgotten that if we wish to commence the study of canal operation starting with an incorrect hydraulic solution, the program will abort indicating the data incompatibility.

The International Commission on Irrigation and Drainage recently published a leaflet entitled "Canal Operation Simulation Models", which lists nineteen variable regime models for canals, among which is the Canvar program we produced, which is mentioned in the following section.

20.6 Operational example of a mathematical model

In order to provide the reader with an idea of what can be accomplished using a model, we have included a study of the first section of the Zújar Canal, for which consideration is being given to the possibility of constructing regulation reservoirs in order to make the desired operational law for the feeding of water into the canal via a hydroelectric plant compatible with the turbine law recommended by the demand for electricity consumption, with the committed law for the delivery of flow rates to the irrigators downstream, which could be any, but in the example is a continuous, flow rate.

The canal capacity is $27.2\,\mathrm{m}^3/\mathrm{s}$, but the necessary irrigation flow rate is $13.6\,\mathrm{m}^3/\mathrm{s}$ continuous. The desirable turbine law is $27.2\,\mathrm{m}^3/\mathrm{s}$ during 12 hours and a practically zero flow rate (very small) during the other 12 hours. The actual flow rate can never be zero, because this is not tolerated by the Saint-Venant equations. Moreover, in practice there will always be water leaking through gates etc., which would make this a real situation.

The canal's transverse section has an arc of circular shape, which only presents the problem where the computer program has to perform calculations that are rather more tedious than those for a trapezoid section.

The purpose of this study is to establish the volumes for the two reservoirs to be installed in the canal, including electric regulation gates at the exits in order to permit the stated turbine law at the entry to deliver a continuous flow rate to the irrigators downstream.

The calculations were performed with the Canvar program (68), (69), (71), which provides the results by means of several outputs.

One form consists of hard-copy printed listings, which are required when the exact results are necessary, however, these are very bulky and inconvenient to handle.

Another form employs computer files, one for flow rates and another for water heights at all points considered in the canal, for each moment of time we are interested in establishing and conserving. With a simple auxiliary program, the data can be used to produce level or flow rate curves by means of a plotter or a graphic printer.

A third data output method from the Canvar program is that of displaying the required curves on the computer screen, which is the fastest and also the most intuitive. Only those curves corresponding to required time periods can be displayed, in which case they are displayed in different colours so that they are easily distinguished, or by displaying the calculated intervals, in which each curve is erased before the following one is drawn. Because of the program and computer operation speed, the results are displayed in the form of a video in which the level curve is seen to vary in accordance with the water movement.

Figures 20.1, 20.2 and 20.3 show the level curves every 2 hours, including the levels of the two reservoirs, checking that after them, in the canal, the

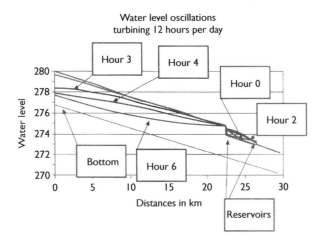

Figure 20.1 Water level oscillations turbining 12 hours a day.

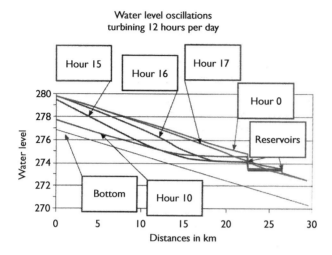

Figure 20.2 Water level oscillations turbining 12 hours a day.

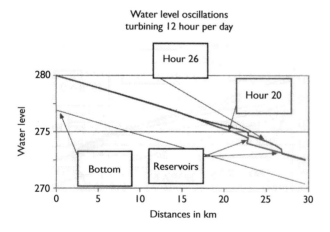

Figure 20.3 Water level oscillations turbining 12 hours a day.

levels are parallel to the bottom and with all the curves superposed, where only one colour is seen because each displayed curve erases the previous one, in other words a constant flow rate has been achieved.

The reservoir volumes were obtained by estimation until the desired values were obtained, which was, in fact, quite easy, due to the program calculation speed. The final volumes were $300\,000$ and $350\,000\,\mathrm{m^3}$.

20.7 Computer applications for canal network design

In general, irrigation canals are not individual, but, instead, combined with other derived ones, forming a network or system. The associated mathematical model can be produced using the same bases and methods, however, it does turn out to be more complex.

If flow rate regulation for the system is carried out upstream, then the process can be simplified. At any moment of time during the study, there will be a certain water demand in each derived canal. This will allow the main canal model to be produced so that it includes the flow rate laws for the secondary canal turnouts. Independently of this, models are produced for each of the other canals, taking into account, in turn, the flow rate laws for each of their associated turnouts.

If the canal system is to be controlled from downstream, the process is still the same as for a single canal, but, again, is more complex from a mathematical point of view. Since the flow rates feeding the derived canals are not known beforehand, all the equations referring to the various sections of each canal must be implemented at each calculation interval. The two equations that define the corresponding gate must be established at the secondary canal turnout points.

The number of equations forming the system to be resolved is therefore much higher than in the case of a single canal, but the employment of modern PCs can overcome any associated difficulties.

Once this equation system, in which the differentials have been replaced by finite increments, has been resolved, the results are introduced into the equation system corresponding to the following interval and so on.

The results produced by the model can then be employed to manage the canal or associated canal network.

Chapter 21

Canal rehabilitation and modernization

21.1 Introduction

Many engineers enjoy designing and building construction works, however, they tend not to be so keen on their later maintenance and repair work. It is a very pleasant thing to see one of our ideas transformed into a physical reality. However, these construction works, especially canals, deteriorate over time and, not only is it necessary to maintain them in good states of repair, but it is also frequently necessary to anticipate this in maintenance operations before it becomes too late, or to modernize them before they become obsolete.

Several words have recently come into fashion among hydraulic engineers, which require a great deal of attention. Some of these are rehabilitation, modernization, maintenance and conservation.

The separation between them is not quite clear, but there should be no doubt that all refer to the fact that hydraulic works should not be left out in the field without any attention being paid to them in order to prevent their deterioration or becoming obsolete. This would mean that the investments put into them are not longer profitable.

For us, and this is how we intend to consider it in this book, the term "rehabilitation" refers to an intervention with respect to a deteriorated construction that requires further work in order to update its services. This may be thought of as being the execution of certain works, not from zero, but commencing with an old construction having substantial problems. Such a case would be a canal with unacceptable filtrations requiring impermeabilization with a new lining, etc.

"Modernization" refers to the modification of significant canal elements that are necessary to confront the new challenges deriving from modern techniques and economy. This would be the case of converting a manually regulated canal into one having automatic control, so that it would be in line with new socio-economic conditioning factors.

However, and again with respect to ourselves, the terms "maintenance" and "conservation" do not involve any substantial differences and refer to the normal operations that have to be periodically carried out to maintain the canal in good conditions and to be able to anticipate deterioration.

Maintenance and conservation are therefore essential operations, in spite of the fact that the governments and other agencies that are responsible for these works are often slow in coming out with the required budgets.

In this chapter, we are going to examine repair and modernization, and in Chapter 22, we will deal with maintenance and conservation.

21.2   Repairing canals

Repair works refer to significant correction operations carried out on deteriorated constructions or those that can no longer be employed for their original purpose. The most frequent and important cases occur when the canal has lost its level of impermeability in a serious fashion.

Figure 21.1 shows the Pisuerga Canal in Spain, with its concrete lining completely deteriorated due to several causes including:

- Poor concrete quality due to insufficient cement (it was constructed between wars)
- Very cold winter weather, which froze the lining surface
- Excessively steep 1/1 slopes that should have been 3/2
- Ground thrust, susceptible to moisture variations
- Insufficient drainage.

Figure 21.1 Pisuerga Canal (Spain).

In such cases, repair work is very difficult because, in addition to other great problems, the work requires the widening of the berm, which would be difficult to achieve because of the high costs of the surrounding land.

The necessary repair work could even require the construction of a new canal; however, the actual repair job will depend on the financial possibilities.

Fortunately, canal repair works are not usually so drastic in scope, but they frequently require a degree of impermeability that they never had previously. This subject has already been indirectly covered at various points in this book. We should recall the usefulness of employing plastic membranes on top of the concrete lining, once it has been established that the latter is of sufficient quality to serve as the foundations (Chapter 8). Here, reference should be made to the Aragón Imperial Canal (Chapter 12), which was given a new lining because it had lost its original one of improved soil over the passage of the years.

It frequently occurs that the impermeable joint material has also lost its qualities and has to be completely replaced. Periodic maintenance operations should be performed to prevent this situation from ever happening and this will be examined in Chapter 22.

Slope slippage can also be quite common, either of the canal itself or of those of the banks. Figure 21.2 shows an example of such slippage of a canal on the Santa Elena peninsula in Ecuador. The problem in this case was caused by the moisture levels in the bank soil.

Figure 21.2 Slope slippage of a canal in Santa Elena (Ecuador).

Cases such as these demonstrate that unsuitable slopes were adopted and it is recommended that not only the affected slopes, but also all others that might be in similar conditions, are repaired before the actual damage occurs. The necessary work itself will vary from case to case and could vary from reducing steep slopes and improving drainage to the construction of thrust-resistant walls, etc.

21.3 Modernizing irrigation canals

In Section 19.1, we described the current situation in irrigation techniques, in which new systems of distributing the water over the land were employed, together with new conditioning factors for the canal. The new socio-economic situation also imposed new types of canal operation, including on-demand irrigation, irrigation periods of much less than 24 hours per day, usually between 12 and 16, depending on the latitude, together with others. Most of these involved the need to transport the amount of water required for one day's irrigation in a time of much less than 24 hours, and this naturally without wasting any unused flow rate at night, employing all other necessary regulations and technical innovations.

When dealing with a new canal, the design conditions are already more or less complicated, but the problem can be resolved with what has been described in this book. The most difficult situation presents itself when an old canal has to be adapted to the new situation. This is a very current problem with maximum economic importance when thought is given to the very large number of canals that have been constructed throughout the world for continuous 24-hour irrigation purposes using a turn system.

The great difficulty involved in this problem can be seen right from the very beginning. If the canal is dimensioned to carry the maximum amount of water that is required for the irrigation zone in one day through continuous 24-hour operation, then it is impossible for this same canal to carry the same volume of water under a variable flow rate law which, at certain times, would exceed the canal's water carrying capacity. This is not a regulation problem, but instead one of a lack of canal capacity.

One solution that has been adopted by Spanish irrigators over many centuries consists of the construction of storage reservoirs at the canal's lateral turnouts to the plots, which would store the delivered water, which would then be employed by the irrigator as necessary. Many areas of Valencia and Murcia, etc., operate in this manner. The system's main drawback is that it occupies a significant area of irrigatable land, which involves far from negligible investment costs. It is obviously a solution, but one that is definitely not the best approach for canal modernization and will not be taken into consideration in this book.

Increasing the canal's size in order to increase its capacity is also undoubtedly another possibility. However, the capacity gains provided by this size increase are invariably much less than that actually required, except in very specific cases, so that this solution is not taken into account either.

Therefore, if we are unable to increase canal capacity, another possibility is to achieve a reduction in flow rates. This is only possible within a framework of the irrigation concept, modernization (67).

It is, of course, quite possible, and even recommendable, that this modernization process takes advantage of the possibility to abandon the cultivation of marginal areas. However, the one real possibility that will enable us to reduce the flow rates through the canal is the actual modernization of the system delivering water to the land. As we explained in Chapter 2, water losses here can reach some 30% of all irrigation water employed throughout the world.

Canal modernization comes from the need to serve new modern irrigation requirements, which include shorter irrigation hours and flexibility in their choice, together with new methods of distributing the water, such as the sprinkler and localized irrigation methods. These can lead to significant savings in water, which is what will enable us to modernize the canal, in addition to the improvements we will carry out on the actual canal itself to reduce filtrations, such as improving the lining, and the fight against unacceptable losses by improving the regulation system.

If we accept, in a simple form, that described in Section 2.1, where one-third of the water entering its head is lost in the canal itself and that another third is lost in the plots then only the final third is actually employed in irrigation. If we also assume that sprinkler irrigation only consumes 70% of that consumed by surface irrigation and that drip irrigation only consumes 50% of the latter, then the changeover from surface irrigation to localized irrigation means dropping from 66% of the water delivered to the plot to only 33%, with a saving of 33% of the total. If the losses in the canal itself can be reduced from 33 to 10%, then we will have achieved a total saving of 56%(33+23%) and this would enable us to provide sufficient water to irrigate the plots in approximately half the time, or in only twelve or thirteen hours per day.

Of course, these values are not absolute and are for guideline purposes only, with each specific case requiring its own calculations to determine those that could be actually attained. Moreover, there will be a very wide range of possible combinations, with only a part being converted to localized irrigation, another to sprinkler and even maintaining surface irrigation in some areas. Each individual case will produce a possible reduction in the canal's circulating flow rate, which will be the final indication of the canal's irrigation flexibility.

However, it is important to stress that it is in the improvement of the plots' irrigation methods and the reduction of water losses where the possibilities will be found to improve the canal's operation and to obtain a much more flexible operability.

In order to achieve this under much more variable and difficult circumstances, many of the regulation elements described in Chapter 19 will be required. All these may be placed within a single concept: to have regulation volumes available, which are either reservoirs connected to the canal, or canal sections that are used simultaneously as the means of water transport and storage through the use of special gates.

Unfortunately, the use of the canal to store water is not always sufficient and will have to be augmented by separate reservoirs. But, in any case, the gates will always fully comply with the aim of providing adequate levels at the lateral turnouts, provided they are correctly chosen and dimensioned.

The location of automatic gates in an old canal for the purpose of improving regulation may come with several problems. The most evident is that that canal was not designed to accommodate the head losses associated with the gates. This will then require either an increase in the canal bank, upstream from each gate, in order to admit a backwater profile, together with the corresponding difference in hydraulic levels or to employ part of the existing freeboard for this purpose. In order to resolve this problem, the possible reduction in circulating flow rates, we have already mentioned, is very important because of the variations in the irrigation systems.

21.4 Modernizing tertiary irrigation canals

As we have already stated, canal modernization requires the updating of the methods of delivering the water to the plots. Between these two elements, canal and plots, we will find the tertiary irrigation canals, which deserve special attention.

The tertiary irrigation canals are elements to which less attention is paid, because of their smaller size, than that given to large main canals. However, their combined losses can become very significant within an irrigatable area because of their total length, and the number of joints may be quite enormous.

The modern trend is towards their replacement by piping, providing this is economically feasible. These are much more adaptable to the demands of the new systems for delivering the water to the plots, by either sprinkler or localized irrigation and, in addition, the water losses per linear metre of piping are significantly lower than those in the tertiary irrigation canals, mainly because of the better quality of the joints.

Commercial piping has such a range of available diameters that they are clear competitors for the tertiary irrigation canals. However, above a certain diameter, they are no longer competitive. For example, above a diameter of 60 cm, the concrete piping has to be reinforced in order to withstand the bending forces due to the weight of the soil and moving overloads. Costs increase with diameters and attention must be paid to the fact that the hydraulic radius in piping is less favourable than in tertiary irrigation canals, because in the latter the free water surface does not have any friction with the walls. This means that in the case of piping, in order to have the same head loss as the tertiary irrigation canals, they have to accept lesser speeds, which entails additional costs.

Above certain diameters, piping is not commercially available, except on special order for large pipelines, and so choice leans towards the use of canals.

We can therefore sum up by saying that the trend is to replace small tertiary irrigation canals with piping up to a certain size, even when taking into account their increased costs. However, for larger sizes, open section canals are quite irreplaceable. An economic study is required in each case and circumstance in order to determine the separation size between watercourses.

Chapter 22

Maintenance and conservation of canals

22.1 Definition

In a paper presented at the International Commission for Irrigation and Drainage, William Price defines canal maintenance as follows:

> Maintenance is the process of keeping existing irrigation, drainage and associated facilities in good repair and good working condition so that all parts of the project's facilities can fulfil the intended purpose for which they were originally designed. It may also cover minor improvements undertaken during the normal process of performing these activities.

Maintenance is a process that can and must be assessed in economic terms because, in the end, it is a cost that can avoid or reduce other costs, possibly much greater costs, whether repairs or replacement of elements forming the installation or production losses. This is why, nowadays, ever-increasing importance is placed on this process and the trend is towards analysing and systematizing it from both the engineering and the economic points of view.

22.2 Characteristics of canals

Canals are basically public works that include a certain number of metal elements, mainly gates and valves, together with the modern items for control and regulation, which can be both complex and delicate because they require sensors, gate drive motors, PLCs, data transmission installations via cable and radio, etc., solar energy panels, electric power lines and buildings to house control and management offices, computers for administrative and technical purposes, etc.

Canal networks are frequently connected to pumping installations, for example sprinkler or drip irrigation or irrigation zones on high ground, etc. All this needs special, specific, very thorough maintenance; however, since this book is limited to the study of canals, they are not included in our study and we shall only examine those mentioned in the previous paragraph. Electric power lines, which normally do not depend on the same organizations that look after canals, are also excluded.

All in all, these elements to be examined are very varied and heterogeneous, which will require maintenance to be carried out by various teams, each with its own training and experience. Nor should it be forgotten that one characteristic of a canal is its long length, which will require transport for the various teams.

Delicate elements supplied by specialized firms, such as sensors and PLCs, will require strict compliance with the manufacturers' instructions. With respect to civil works, it is necessary to commence with good construction engineering practices, together with the logical use of experience gained over years of operation.

The organization that manages the canal should be aware of the fact that maintenance is a costly, but essential operation that is profitable when correctly carried out since it can prevent much higher costs. During maintenance planning studies, it is recommended to have an idea of the probable lifetimes for the various elements in the installations because this will provide some indication of their replacement costs. Without any desire to convince the reader to adopt the figures given in many publications, we would suggest the following figures:

Lifetime (years)

Canals, roadways and buildings	75
Gates, valves and closed pipelines	40
Measurement and telecommunications equipment, etc.	10

There is currently a great deal of emphasis on maintenance problems, but this is mainly aimed at pumping and electromechanical installations, with a lot less for civil works. At least this is the impression gained from congresses and technical journals. Overall, this is a matter requiring much analysis and perfection and this book is only going to sketch out certain important ideas.

22.3   Types of maintenance

There are several types of maintenance, broadly classified under:

1   Preventive maintenance
2   Corrective maintenance.

Preventive maintenance is carried out on a routine basis before any failure or defect is discovered in the installations and one attempts to reduce the probability of their presentation or problems with the service.

It is usually carried out in accordance with a pre-established programme that may be planned over certain time intervals or depending on the appearance of certain indications.

On many occasions the time separating two successive preventive maintenance operations may be chosen and this should be established in accordance with profitability criteria such as benefit/cost.

For purely guideline purposes, we can say that normally one year is allowed to elapse between two preventive maintenance inspections for electromechanical elements, which mainly covers gates, piping and their joints.

Depending on their conditions of use, suspension cables and chains may have inspection periods of between one month and one year.

Accumulator batteries and automatic electricity generators that are employed to provide backup power for gates and data transmission should be inspected once in a month, with a more thorough inspection taking place once a year.

Corrective maintenance, on the other hand, takes place after a problem occurs. The actual problem itself will generally determine the necessary level of urgency and depth of scope. They normally consist of repairs required to maintain or restore service and this objective will form the basis for any decisions to be taken.

Depending on the type of affected installation, certain defective parts may require replacement. This obviously means that some form of spare part store be available, where the parts actually held should be the result of a study using probability calculations based on experience or the manufacturers' manuals that indicate the probable lifetime for a given element or material. Nevertheless, it is always recommendable to adopt a certain safety factor, which should increase with the distance to the canal's location and/or the country's degree of development.

On occasions, the problem may be so complex that the rehabilitation projects described in the previous chapter become necessary. However, it is not a good idea to allow this type of situation to arise, because it could indicate that a poor level of maintenance management is actually in place.

22.4 Maintenance and conservation of civil works

In order to achieve good levels of maintenance and conservation in canals, whatever their purpose, it is essential for the personnel to carry out frequent inspections under the careful supervision of the chief engineer responsible for the canal, so that any emerging problem can be detected at a very early stage of development.

However, certain points, due to frequency or importance, require special attention. These are vegetation growth control, the early detection of filtrations, sediment removal and the conservation of metal elements.

22.4.1 Vegetation control

We have already stated that aquatic vegetation is liable to proliferate inside a canal. This is due to the availability of large quantities of water, bottom sediment for root growth and, especially with respect to irrigation canals, abundant sunlight and good temperatures, all of which is highly favourable to prolific vegetation growth.

The Imperial Aragón Canal, which was constructed two centuries ago, over permeable gravel ground, with an improved soil lining, has suffered such a reduction of its lining due to the very necessary cleaning operations to remove the luxuriant vegetation growing continually inside that several solutions had to be adopted to renew its lining. This is an example of the damage that can be caused by vegetation, mainly in unlined canals or those with lined with soil.

Nevertheless, there is no other solution but to periodically remove it. Section 10.3 described how vegetation is one of the reasons for freeboard. Cleaning operations were traditionally performed by hand, taking advantage of any periods when the canal is empty. Vegetation has also been frequently eliminated by burning at appropriate times; with this often being carried out in two phases: the first for drying purposes with a flamethrower, followed by a second final operation (54). Nowadays there is the possibility of fitting cutting blades to backhoes, which are then employed to "shave" the growing plants. It should not be necessary to explain that the operator must exercise maximum care and skill not to go very deep into the soil, which could lead to an unwanted increase in the canal's transverse section, with the associated inconveniences of reducing the thickness of an improved soil lining. Another mechanical method for eliminating vegetation is that of dragging a heavy chain along the canal bottom between two tractors, one on each side.

On many occasions, however, the problem involves floating vegetation that is capable of blocking gate mechanisms and obstructing sensors, etc. One solution here is to install gratings or nets across the canal that can intercept the passage of this vegetation, but which will also require frequent cleaning.

Even rigidly lined canals, mainly those with concrete linings, do not escape the problems associated with vegetation. Whereas it is true that the cleaning of these canals is much simpler, and is one of the reasons for lining a canal, vegetation will still grow inside them.

Figure 22.1 shows how vegetation can grow and damage a joint, raising and deteriorating a concrete lining, and also Figure 5.2 demonstrates the ability of plants to take root even in such unfavourable sites as the concrete lining itself.

Rapid repair of these problems will prevent even greater ones appearing in the near future.

Figure 22.1 Vegetation in Huaral Canal (Peru).

If the canal is lined with plastic sheeting or bituminous products, then the growth of vegetation will be one of its greatest enemies and it is essential to minimize any damage right from the very beginning.

All canals should undergo systematic inspections on an annual basis, irrigation canals at the end of the season, even if it does not appear to be necessary, with all maintenance and repair operations being strictly carried out.

The problem with vegetation growing canals is so great that much effort has gone into finding chemical products that can be added to the circulating water to reduce their growth, but without causing any damage to crops or endangering human consumption. We are of the opinion that, to date, no such products have been found that fully comply with these requirements.

A joint study by the North American Department of Agriculture and the Bureau of Reclamation (23) mentions aromatic solvents, acrolein and copper sulphate as some of the analysed products.

Aromatic solvents are hydrocarbons with molecules containing hexagonal, benzene-like bonds. Since they are not soluble in water, they require prior mixing and preparation with an emulsifier, which causes the product to form an emulsion that disperses in the water instead of tending to float.

Canal vegetation that can be combated with these products is very diverse and will naturally vary with the natural conditions of the area. The cited North American study also describes how agricultural crops have been obtained without any appreciable reduction in production. Among these crops are found alfalfa, carrots, cotton, sorghum, lettuce, wheat, barley, meadows, potatoes, sugar beet and sweet corn.

However, the study also stated that aromatic solvents are highly toxic to fish, snails and insects, which will die if they come into contact with the treated water. Livestock does not like the taste and will not normally drink water containing it. Finally, these solvents are highly inflammable and require extreme precaution when handling them.

Acrolein does not appear to be very tempting either. It is highly toxic to mammals, fish, frogs, snails and other aquatic organisms. It is a product that causes irritation to the eyes and respiratory tract. Moreover, it is very dangerous to handle because it reacts violently with both acids and alkalis. In its favour, the report states that almost all aquatic vegetation is sensitive to acrolein, and the advantage over aromatic solvents is that it is soluble in water.

Perhaps the only usable product of the three studied is copper sulphate, which is very efficient in the fight against algae. However, it is capable of killing fish, but in the correct proportion it should not harm either humans or livestock.

We should stress, at this point, that the use of herbicides in canals is prohibited in some countries.

We will sum up by stating that the problem of vegetation inside canals can be very significant and usually requires careful, periodic cleaning operations.

22.4.2 Early detection of filtration

When a canal begins to suffer from infiltrations, these will drag along the fine material from the land and may cause dangerous undercutting, and if the canal is lined it may produce cavities in the lining. These will increase in size, leading to lining rupture, with a heavy flow of water that could cause the canal backfill to collapse with consequent flooding of the irrigatable area. One might label us as alarmists, but when a person has lived through such an experience it is not so easily forgotten.

Frequent inspections to detect possible filtrations in canals form a very important task. If there is any drainage underneath the canal bottom, the outlets should be inspected, checking for increased drainage flow rates, even carrying out regular measurements to establish any variation.

Whether drainage exists or not, it is necessary to frequently inspect the contact edge of backfills with the natural land, looking for not only pools or wet areas, but also the growth of plants that search out moisture and which can be easily detected on arid and semi-arid land.

We have worked on a canal with filtrations that caused the surrounding groundwater to rise, but without any external evidence other than the problems in the nearby cultivated areas.

In this book, we have stressed the need to fight against filtrations in order not to waste water. Now we are going to also point out the importance that filtrations can have with respect to the canal's actual stability.

Filtration repairs should be carried out as soon as possible, but there are times when it is not possible to shut down canal service. When describing plastic sheeting, we explained its usefulness for the temporary repair of concrete-lined canals that have cracked, in order to seal off water leaks, at least until the end of the irrigation season when permanent repairs may be carried out. In these cases, the membrane should be both light and manageable, with PVC being the ideal material.

Periodic inspections followed by any necessary repairs of joints in concrete-lined canals is an essential task. We have already described how materials employed to impermeabilize canals can rapidly age, losing elasticity, peel away from the concrete and produce filtrations.

22.4.3 *Cleaning up sediments*

Solids carried along by the water will nearly always sediment out in the same locations in a canal. These are the solid particles that the water flow is no longer capable of carrying along.

A typical location is at inverted syphon pipes. The water, as it leaves the pipe and begins its upward journey through the syphon outlet, is unable to carry the particles with it due to the force of gravity and hence sediment collects at the bottom. The actual problem is not serious as long as the deposits do not grow sufficiently to produce significant head loss. Although the sediment is removed when the canal shows signs of becoming sluggish, the problem will continue, because the replacement of such sediment can occur very quickly.

In fact, what is really serious is sedimentation in small watercourses, which then can become blocked. What has to be achieved is that the water always has to possess sufficient tractive force to prevent sedimentation, but, of course, without causing any erosion in the case of an unlined canal. This is not an easy problem to solve and, on many occasions, the only solution is to install settling tanks (Section 11.8), for which it is not always possible to find cleaning outlets. We have already discussed the beneficial effect of regulation reservoirs installed in the canal. Both require periodic cleaning operations.

22.5 Conservation of metal elements

Metal canal elements are mainly the gates and also lateral syphon spillways and gratings. Sometimes there are also important syphons, as the Dehesilla Syphon in Orellana Canal (Figure 22.2). Except for certain special elements, when we speak of metal elements in civil works, we invariably mean iron and steel.

Figure 22.2 Dehesilla Syphon in Orellana Canal.

These elements are found either submerged in water or in very damp environments, which is highly unfavourable to their maintenance.

For good gate maintenance, it is essential in their manufacture that such elements undergo careful descaling before they are painted.

When the metal sheet is rolled, slag or points having a different chemical composition can become embedded, and form what is known as "scale". This can have thermal dilation and elastic deformation coefficients that are different from the rest of the material. During the iron element's lifetime in the canal, the scale is deformed in a different fashion, which causes it to peel away from the metal sheet. The problem now occurs when the elements have been protected by paint, because this will also peel away with the scale and leave unprotected areas to rapidly suffer oxidation or rusting. There is also the additional problem when the scale is not the same material as metal sheet, which can lead to galvanic corrosion. In this process, the water dissolves certain salts forming an electrolyte and hence an electric cell in which the corrodible pole is the iron, which again undergoes oxidation.

This is why a full descaling operation is essential, by means of either sand or shot blasting, before the first primer coat. The employed air pressure should be approximately seven atmospheres.

The paint that can be employed is available in several specialized treatments (22), providing many possibilities with the different types.

It is recommended that the first time the gate is painted, it be given two primer coats of a protection paint known as "wash primer". This, in fact, is a paint that contains metals that are below iron on the galvanic corrosion scale, such as zinc or aluminium. This will provide the guarantee that they will oxidize before the gate material. The paint solvent is light and fluid so that the each coat is only a few microns in thickness, for example, 10 microns per coat.

Two coats of a good paint should be applied on top of this. Vinyl paints are usually used for gates, with bitumen-based paint for the gratings.

Some gate manufacturers apply the following paint process at the factory, after a thorough descaling operation: one coat of anti-corrosion epoxy bitumen shop primer, with a 40-micron thickness, followed by two coats of epoxy-tar, formulated with epoxy resins and bitumen.

If protecting metal elements to be installed outside, but not submerged in water, then the following procedure may be adopted: one primer coat of chlororubber red lead, with a thickness of 30 microns, followed by two coats of chlororubber topcoat, each with a thickness of 30 microns.

The best and also the most expensive method consists of zincing the parts. This may be carried out at the factory either by electrolysis or by immersion in molten zinc after a thorough descaling process. However, this is not usually employed for gates. When zinc protection is required, this is normally applied by spraying with a gun, with a thickness of 80 microns. This is then followed by a 30-micron thick coat of epoxy paint.

Assuming that all these basic instructions have been followed in the construction of the gates, then during the element's lifetime, it will be necessary to regularly and thoroughly inspect them so that when necessary the paintwork can be repaired before it becomes too late. If this is always carried out in this way, then most repair work will be small scale, otherwise a complete repainting may be required after a further thorough descaling process. In the latter case, instead of transporting the gate to a workshop and sand-blasting, the descaling operation may be performed on site using a rotating electric brush, since manual operation would not suffice. Obviously some form of generator will also be required to power the brush.

It is recommended that the paints employed in any repair work be the same as those used for the very first time in order to avoid any incompatibilities between the various types of paint.

However, certain delicate elements, some of which may be made of metal, must be protected from rough handling and vandalism by persons with easy access to them. This will require the construction of works to provide some degree of protection.

One very clear case of this is the sensor equipment for the measurement of levels, flow rates, gate apertures and PLCs, etc. All these must be installed inside some form of concrete structure that provides them with reasonable protection.

Even automatic gates can suffer the unwanted attention of vandals. These are therefore occasionally protected inside a metal mesh structure which, although it may not provide a great deal of physical protection, does at least clearly define private property that should be left alone.

Chapter 23

Canal safety

In spite of their often apparently tranquil, serene appearance, canals can involve great danger for human beings. A person falling into a canal or one who freely enters the water to swim quite probably may not be able to get out. The canal sides, which are slippy and have a greater inclination than the victim would expect, do not have any protrusions that can be grasped by desperate hands, especially if it is lined, all of which make it very difficult to get out. This is the situation with all canals, but when dealing with canals employed in hydroelectric schemes, with their faster flow rates than those for irrigation, this danger greatly increases.

Protection for the general public should begin with making it fully aware that canals are dangerous, and should only be approached by those persons who are responsible for its maintenance or monitoring, etc. The signs indicating that it is strictly forbidden to bathe in canals should never be ignored and it is strongly recommended that the public stays well away from them.

Walls, fences and other structures should be employed to prevent access to all especially dangerous areas, such as rapids, tunnels, flumes and syphons. In such locations, anybody being dragged along by the current in areas where this is stronger than normal has very little probability of being able to dominate the situation. Figure 23.1 shows the fenced-off area at the entrance to Tunnel 2 on the Orellana Canal.

Syphons are especially dangerous to life and limb, where a person may be swept inside the pipe. Under these conditions it is impossible to breathe and in addition, severe injury can result from contact with the walls. There have been cases in which, because the pipe is very narrow, the victim becomes trapped inside and is only detected when the canal overflows.

All this requires a series of safety measures, which should include a minimum of those already described and other essential ones.

One measure is to make it possible for the person to be able to get out without the help of other persons, since calls for help are not often heard.

The required minimum is to install metal ladders along the length of the canal. These should be 40 cm wide consisting of round 12 mm diameter bars that are separated by a maximum of 40 cm. These ladders may be

Figure 23.1 Fenced-off area at the entrance to Tunnel 2 on the Orellana Canal.

constructed in a single piece, with metal side profiles fixed to the walls, or, when the canal is concrete lined, the individual rungs can be embedded in the concrete. These ladders should be installed approximately every 200 m along the canal, alternatively on the two sides and always before a danger area. Figure 23.2 shows a step ladder at the exit of the Talave Canal tunnel, and another on the Chongón-Subeybaja Canal in Ecuador in Figure 8.5.

The purpose of these ladders is to allow possible victims to let the flowing water carry them along to the nearest ladder.

A system is regularly employed in the USA, in which ropes are stretched over the canal from side to side with hanging leather straps that a victim may use to get out (Figure 23.3).

Because syphons are so dangerous, extreme precautions should be employed at their entrances. It is recommended that metal or nylon nets are installed in the canal from one side to the other. Grilles are sometime fitted to the syphon entrance to prevent foreign material blocking them up inside. These grilles also form an excellent life-saving element, provided the bars are sufficiently angled to help people to get out. At the same time, they allow scrapers to be used to remove vegetation and rubbish.

The safety of human life is not only achieved using protection against possible drowning situations, but should also include health protection measures. The dumping of rubbish from relatively close population centres are frequent dangers in certain locations, which require constant police

Figure 23.2 Step ladder at the exit of the Talave Canal tunnel.

Figure 23.3 Ropes stretched over the canal from side to side with hanging leather straps.

vigilance. On occasions, the only remedy is to cover the canal and this may be essential in canals supplying drinking water. The Canal Alto on the Isabel II Canal (Madrid) is completely covered by a dome for this very reason, with the same applying to the Bajo Canal and the Atazar Canal (both in Madrid).

The dumping of chemical waste products from factories, etc. is especially dangerous to health and not just in drinking water supplies, but also in irrigation canals. In these situations, strict control measures must be put into operation.

Discharge of standard contracted rectangular weirs (in m³/s) (Francis Formula)

Head H (cm)	Length of weir L (cm)								
	15.00	25.00	50.00	75.00	100.00	125.00	150.00	175.00	200.00
0.50	.0001	.0002	.0003	.0005	.0006	.0008	.0009	.0011	.0013
1.00	.0003	.0005	.0009	.0014	.0018	.0022	.0027	.0032	.0036
1.50	.0005	.0008	.0017	.0025	.0033	.0042	.0050	.0059	.0067
2.00	.0008	.0013	.0026	.0039	.0051	.0064	.0077	.0090	.0103
2.50	.0011	.0018	.0036	.0054	.0072	.0090	.0108	.0126	.0145
3.00	.0014	.0023	.0047	.0071	.0095	.0118	.0142	.0166	.0190
3.50	.0017	.0029	.0059	.0089	.0119	.0149	.0179	.0209	.0239
4.00	.0021	.0036	.0072	.0109	.0145	.0182	.0219	.0256	.0293
4.50	.0025	.0042	.0086	.0130	.0173	.0217	.0261	.0305	.0349
5.00	.0029	.0049	.0101	.0152	.0203	.0254	.0306	.0357	.0409
5.50	.0033	.0057	.0116	.0175	.0234	.0293	.0353	.0412	.0471
6.00	.0037	.0064	.0132	.0199	.0267	.0334	.0402	.0469	.0537
6.50	.0042	.0072	.0148	.0224	.0300	.0376	.0453	.0529	.0605
7.00	.0046	.0080	.0166	.0251	.0335	.0420	.0506	.0591	.0676
7.50	.0051	.0089	.0183	.0277	.0371	.0466	.0560	.0655	.0749
8.00		.0097	.0201	.0305	.0409	.0513	.0617	.0721	.0825
8.50		.0106	.0220	.0334	.0447	.0561	.0675	.0789	.0903
9.00		.0115	.0239	.0363	.0487	.0611	.0735	.0859	.0983
9.50		.0124	.0259	.0393	.0528	.0662	.0797	.0931	.1066
10.00		.0134	.0279	.0424	.0569	.0715	.0860	.1005	.1151
10.50		.0143	.0300	.0456	.0612	.0768	.0925	.1081	.1237
11.00		.0153	.0321	.0488	.0656	.0823	.0991	.1159	.1326
11.50		.0163	.0342	.0521	.0700	.0879	.1059	.1238	.1417
12.00		.0173	.0364	.0555	.0745	.0936	.1128	.1319	.1510
12.50		.0183	.0386	.0589	.0792	.0995	.1198	.1401	.1604
13.00		.0193	.0409	.0624	.0839	.1054	.1270	.1485	.1701
13.50		.0203	.0431	.0659	.0887	.1115	.1343	.1571	.1799
14.00		.0214	.0455	.0695	.0936	.1176	.1417	.1658	.1899
14.50		.0224	.0478	.0732	.0985	.1239	.1493	.1747	.2000
15.00		.0235	.0502	.0769	.1036	.1303	.1570	.1837	.2104
15.50		.0246	.0526	.0807	.1087	.1367	.1648	.1928	.2209
16.00			.0551	.0845	.1139	.1433	.1727	.2021	.2315

16.50	.0575	.0883	.1191	.1499	.1807	.2115	.2423
17.00	.0601	.0923	.1244	.1567	.1889	.2211	.2533
17.50	.0626	.0962	.1298	.1635	.1971	.2308	.2644
18.00	.0651	.1002	.1353	.1704	.2055	.2406	.2757
18.50	.0677	.1043	.1408	.1774	.2140	.2505	.2871
19.00	.0703	.1084	.1464	.1845	.2226	.2606	.2987
19.50	.0730	.1126	.1521	.1917	.2312	.2708	.3104
20.00	.0756	.1167	.1578	.1989	.2400	.2811	.3222
20.50	.0783	.1210	.1636	.2063	.2489	.2916	.3342
21.00	.0810	.1253	.1694	.2137	.2579	.3021	.3464
21.50	.0838	.1296	.1754	.2212	.2670	.3128	.586
22.00	.0865	.1339	.1813	.2287	.2762	.3236	.3710
22.50	.0893	.1383	.1873	.2364	.2854	.3345	.3835
23.00	.0921	.1428	.1934	.2441	.2948	.3455	.3962
23.50		.1472	.1995	.2519	.3043	.3566	.4090
24.00		.1517	.2057	.2598	.3138	.3679	.4219
24.50		.1563	.2120	.2677	.3235	.3792	.4349
25.00		.1609	.2183	.2757	.3332	.3906	.4481
25.50		.1655	.2246	.2838	.3430	.4022	.4614
26.00		.1701	.2310	.2919	.3529	.4138	.4747
26.50		.1748	.2375	.3002	.3629	.4256	.4883
27.00		.1795	.2440	.3084	.3729	.4374	.5019
27.50		.1843	.2505	.3168	.3831	.4493	.5156
28.00		.1890	.2571	.3252	.3933	.4614	.5295
28.50		.1938	.2637	.3337	.4036	.4735	.5434
29.00		.1987	.2704	.3422	.4140	.4857	.5575
29.50		.2035	.2771	.3508	.4244	.4981	.5717
30.00		.2084	.2839	.3594	.4350	.5105	.5860
30.50		.2134	.2907	.3682	.4456	.5230	.6004
31.00			.2976	.3769	.4563	.5356	.6149
31.50			.3045	.3858	.4670	.5483	.6295
32.00			.3115	.3946	.4778	.5610	.6442
32.50			.3184	.4036	.4888	.5739	.6591
33.00			.3255	.4126	.4997	.5869	.6740
33.50			.3325	.4217	.5108	.5999	.6890
34.00			.3396	.4308	.5219	.6130	.7041
34.50			.3468	.4399	.5331	.6262	.7193
35.00			.3540	.4492	.5443	.6395	.7347
35.50			.3612	.4584	.5556	.6529	.7501
36.00			.3685	.4677	.5670	.6663	.7656
36.50			.3758	.4771	.5785	.6798	.7812
37.00			.3831	.4865	.5900	.6934	.7969
37.50			.3905	.4960	.6016	.7071	.8127
38.00			.3979	.5055	.6132	.7209	.8285
38.50			.4053	.5151	.6249	.7347	.8445
39.00			.4128	.5247	.6367	.7486	.8606
39.50			.4203	.5344	.6485	.7626	.8767
40.00			.4278	.5441	.6604	.7767	.8929
40.50			.4354	.5539	.6723	.7908	.9093

Head H (cm)	Length of weir L (cm)								
	15.00	25.00	50.00	75.00	100.00	125.00	150.00	175.00	200.00
41.00					.4430	.5637	.6843	.8050	.9257
41.50					.4507	.5735	.6964	.8193	.9422
42.00					.4583	.5834	.7085	.8336	.9587
42.50					.4660	.5934	.7207	.8481	.9754
43.00					.4738	.6034	.7330	.8626	.9921
43.50					.4815	.6134	.7452	.8771	1.0090
44.00					.4893	.6235	.7576	.8917	1.0259
44.50					.4971	.6336	.7700	.9064	1.0429
45.00l					.5050	.6437	.7825	.9212	1.0599
45.50l					.5129	.6539	.7950	.9360	1.0771
46.00					.5208	.6642	.8075	.9509	1.0943
46.50					.5287	.6744	.8202	.9659	1.1116
47.00					.5366	.6847	.8328	.9809	1.1290
47.50					.5446	.6951	.8456	.9960	1.1465
48.00					.5526	.7055	.8583	1.0112	1.1640
48.50					.5607	.7159	.8712	1.0264	1.1816
49.00					.5687	.7264	.8840	1.0417	1.1993
49.50					.5768	.7369	.8970	1.0570	1.2171
50.00					.5849	.7474	.9099	1.0724	1.2349

Appendix 2

Free flow discharge values for Parshal Measuring Flume

H_a (cm)	W (cm)								
	15.25	22.86	30.48	45.72	60.96	91.44	121.92	152.40	182.88
3.00	.0015	.0025	.0033	.0048					
3.50	.0019	.0032	.0042	.0060					
4.00	.0024	.0039	.0052	.0074					
4.50	.0028	.0047	.0062	.0089	.0116	.0169			
5.00	.0034	.0055	.0072	.0105	.0137	.0200			
5.50	.0039	.0063	.0084	.0122	.0159	.0232			
6.00	.0045	.0072	.0096	.0139	.0182	.0266	.0348	.0429	
6.50	.0051	.0082	.0108	.0157	.0206	.0302	.0395	.0487	
7.00	.0057	.0092	.0121	.0176	.0231	.0339	.0444	.0548	
7.50	.0064	.0102	.0134	.0196	.0257	.0378	.0495	.0611	.0726
8.00	.0071	.0112	.0148	.0217	.0285	.0418	.0549	.0677	.0805
8.50	.0078	.0123	.0162	.0238	.0313	.0459	.0604	.0746	.0887
9.00	.0085	.0135	.0177	.0260	.0342	.0503	.0661	.0817	.0971
9.50	.0093	.0146	.0192	.0282	.0372	.0547	.0720	.0890	.1059
10.00	.0100	.0158	.0208	.0306	.0402	.0593	.0780	.0965	.1149
10.50	.0108	.0170	.0224	.0329	.0434	.0640	.0843	.1043	.1242
11.00	.0117	.0183	.0240	.0354	.0466	.0688	.0907	.1123	.1338
11.50	.0125	.0196	.0254	.0379	.0500	.0738	.0973	.1205	.1436
12.00	.0134	.0209	.0274	.0405	.0534	.0789	.1040	.1290	.1537
12.50	.0143	.0222	.0292	.0431	.0569	.0841	.1110	.1376	.1640
13.00	.0152	.0236	.0310	.0458	.0604	.0894	.1181	.1464	.1746
13.50	.0161	.0250	.0328	.0485	.0641	.0949	.1253	.1555	.1854
14.00	.0171	.0264	.0347	.0513	.0678	.1004	.1327	.1647	.1965
14.50	.0180	.0279	.0360	.0541	.0716	.1061	.1403	.1741	.2078
15.00	.0190	.0294	.0385	.0570	.0755	.1119	.1480	.1838	.2194
15.50	.0200	.0309	.0405	.0600	.0794	.1178	.1558	.1936	.2311
16.00	.0211	.0324	.0425	.0630	.0834	.1238	.1638	.2036	.2431
16.50	.0221	.0340	.0445	.0661	.0875	.1299	.1720	.2138	.2554
17.00	.0232	.0356	.0466	.0692	.0916	.1361	.1803	.2242	.2678
17.50	.0243	.0372	.0487	.0723	.0958	.1425	.1887	.2347	.2805

(Continued)

H_a (cm)	W (cm)								
	15.25	22.86	30.48	45.72	60.96	91.44	121.92	152.40	182.88
18.00	.0254	.0388	.0508	.0755	.1001	.1489	.1973	.2455	.2934
18.50	.0265	.0405	.0530	.0788	.1045	.1554	.2060	.2564	.3065
19.00	.0276	.0422	.0552	.0821	.1089	.1620	.2149	.2675	.3198
19.50	.0288	.0439	.0574	.0854	.1133	.1688	.2239	.2787	.3333
20.00	.0300	.0456	.0597	.0888	.1179	.1756	.2330	.2901	.3471
20.50	.0312	.0474	.0619	.0923	.1225	.1825	.2423	.3017	.3610
21.00	.0324	.0492	.0643	.0957	.1271	.1896	.2516	.3135	.3752
21.50	.0336	.0509	.0666	.0993	.1319	.1967	.2612	.3254	.3895
22.00	.0349	.0528	.0690	.1029	.1366	.2039	.2708	.3375	.4040
22.50	.0361	.0546	.0714	.1065	.1415	.2112	.2806	.3498	.4188
23.00	.0374	.0565	.0738	.1101	.1464	.2186	.2905	.3622	.4337
23.50	.0387	.0584	.0762	.1138	.1513	.2261	.3005	.3748	.4489
24.0	.0400	.0603	.0787	.1176	.1564	.2337	.3107	.3875	.4642
24.5	.0413	.0622	.0812	.1214	.1614	.2413	.3210	.4004	.4797
25.0	.0427	.0642	.0838	.1252	.1666	.2491	.3314	.4134	.4954
25.5	.0440	.0661	.0863	.1291	.1718	.2569	.3419	.4267	.5113
26.0	.0454	.0681	.0889	.1330	.1770	.2649	.3525	.4400	.5274
26.5	.0468	.0701	.0915	.1370	.1823	.2729	.3633	.4535	.5436
27.0	.0482	.0722	.0942	.1410	.1877	.2810	.3741	.4672	.5601
27.5	.0496	.0742	.0968	.1450	.1931	.2892	.3851	.4810	.5767
28.0	.05in_	.0763	.0995	.1491	.1986	.2975	.3962	.4949	.5935
28.5	.0525	.0784	.1023	.1532	.2041	.3058	.4075	.5090	.6105
29.00	.0539	.0805	.1050	.1573	.2097	.3143	.4188	.5233	.6277
29.5	.0554	.0826	.1078	.1615	.2153	.3228	.4303	.5377	.6451
30.0	.0569	.0848	.1106	.1658	.2210	.3314	.4418	.5522	.6626
30.5	.0583	.0870	.1134	.1700	.2267	.3401	.4535	.5669	.6803
31.0	.0599	.0892	.1162	.1743	.2325	.3489	.4653	.5817	.6981
31.5	.0614	.0914	.1191	.1787	.2383	.3577	.4772	.5967	.7162
32.0	.0629	.0936	.1219	.1831	.2442	.3667	.4892	.6118	.7344
32.5	.0645	.0959	.1248	.1875	.2502	.3757	.5013	.6270	.7528
33.0	.0661	.0981	.1278	.1919	.2562	.3848	.5135	.6424	.7713
33.5	.0677	.1004	.1307	.1964	.2622	.3939	.5259	.6579	.7901
34.0	.0693	.1027	.1337	.2010	.2683	.4032	.5383	.6736	.8089
34.5	.0709	.1050	.1367	.2055	.2744	.4125	.5508	.6893	.8280
35.0	.0725	.1074	.1398	.2101	.2806	.4219	.5635	.7053	.8472
35.5	.0742	.1097	.1428	.2148	.2869	.4314	.5762	.7213	.8666
36.0	.0758	.1121	.1459	.2194	.2932	.4410	.5891	.7375	.8861
36.5	.0775	.1145	.1490	.2241	.2995	.4506	.6021	.7538	.9058
37.0	.0792	.1169	.1521	.2289	.3059	.4603	.6151	.7703	.9257
37.5	.0809	.1193	.1552	.2337	.3123	.4701	.6283	.7869	.9457
38.0	.0826	.1218	.1584	.2385	.3188	.4799	.6416	.8036	.9659
38.5	.0843	.1242	.1616	.2433	.3253	.4898	.6549	.8204	.9863
39.0	.0861	.1267	.1648	.2482	.3319	.4998	.6684	.8374	1.007
39.5	.0878	.1292	.1680	.2531	.3385	.5099	.6820	.8545	1.027
40.0	.0896	.1317	.1713	.2580	.3452	.5201	.6957	.8718	1.048
40.5	.0914	.1342	.1745	.2630	.3519	.5303	.7094	.8891	1.069

41.0	.0932	.1368	.1778	.2680	.3586	.5406	.7233	.9066	1.090
41.5	.0950	.1394	.1811	.2731	.3654	.5509	.7373	.9242	1.112
42.0	.0968	.1419	.1845	.2782	.3723	.5614	.7513	.9419	1.133
42.5	.0986	.1445	.1878	.2833	.3792	.5719	.7655	.9598	1.155
43.0	.1004	.1471	.1912	.2884	.3861	.5824	.7798	.9778	1.176
43.5	.1023	.1498	.1946	.2936	.3931	.5931	.7941	.9959	1.198
44.0	.1042	.1524	.1980	.2988	.4001	.6038	.8086	1.014	1.220
44.5	.1060	.1551	.2014	.3040	.4072	.6146	.8231	1.033	1.243
45.0	.1079	.1577	.2049	.3093	.4143	.6254	.8377	1.051	1.265
45.5		.1604	.2084	.3146	.4214	.6363	.8525	1.070	1.287
46.0		.1631	.2119	.3199	.4286	.6473	.8673	1.088	1.310
46.5		.1659	.2154	.3253	.4359	.6584	.8822	1.107	1.333
47.0		.1686	.2189	.3307	.4432	.6695	.8972	1.126	1.356
47.5		.1713	.2225	.3361	.4505	.6807	.9124	1.145	1.379
48.00		.1741	.2260	.3416	.4579	.6919	.9276	1.164	1.402
48.50		.1769	.2296	.3471	.4653	.7033	.9428	1.184	1.425
49.00		.1797	.2333	.3526	.4727	.7147	.9582	1.203	1.449
49.50		.1825	.2369	.3581	.4802	.7261	.9737	1.223	1.473
50.00		.1853	.2405	.3637	.4878	.7376	.9893	1.242	1.496
50.50		.1882	.2442	.3693	.4953	.7492	1.005	1.262	1.520
51.00		.1910	.2479	.3750	.5030	.7609	1.021	1.282	1.544
51.50		.1939	.2516	.3806	.5106	.7726	1.037	1.302	1.569
52.00		.1968	.2553	.3863	.5183	.7844	1.052	1.322	1.593
52.50		.1997	.2591	.3921	.5261	.7962	1.068	1.342	1.617
53.00		.2026	.2628	.3978	.5339	.8081	1.085	1.363	1.642
53.50		.2056	.2666	.4036	.5417	.8201	1.101	1.383	1.667
54.00		.2085	.2704	.4094	.5495	.8321	1.117	1.404	1.692
54.50		.2115	.2743	.4153	.5575	.8442	1.133	1.424	1.717
55.00		.2144	.2781	.4212	.5654	.8564	1.150	1.445	1.742
55.50		.2174	.2820	.4271	.5734	.8686	1.166	1.466	1.767
56.00		.2204	.2858	.4330	.5814	.8809	1.183	1.487	1.793
56.50		.2235	.2897	.4390	.5895	.8932	1.200	1.508	1.818
57.00		.2265	.2936	.4449	.5976	.9057	1.217	1.529	1.844
57.50		.2295	.2976	.4510	.6057	.9181	1.233	1.551	1.870
58.00		.2326	.3015	.4570	.6139	.9307	1.250	1.572	1.896
58.50		.2357	.3055	.4631	.6221	.9433	1.267	1.594	1.922
59.00		.2388	.3095	.4692	.6304	.9559	1.285	1.615	1.948
59.50		.2419	.3135	.4753	.6387	.9686	1.302	1.637	1.975
60.00		.2450	.3175	.4815	.6470	.9814	1.319	1.659	2.001
60.50		.2481	.3215	.4877	.6554	.9943	1.336	1.681	2.028
61.00		.2513	.3256	.4939	.6638	1.007	1.354	1.703	2.055
61.50			.3296	.5001	.6723	1.020	1.371	1.725	2.082
62.00			.3337	.5064	.6808	1.033	1.389	1.748	2.109
62.50			.3378	.5127	.6893	1.046	1.407	1.770	2.136
63.00			.3420	.5190	.6878	1.059	1.425	1.793	2.163
63.50			.3461	.5254	.7064	1.073	1.443	1.815	2.191
64.00			.3503	.5317	.7151	1.086	1.460	1.838	2.218
64.50			.3544	.5381	.7238	1.099	1.479	1.861	2.246
65.00			.3586	.5446	.7325	1.113	1.497	1.884	2.274

(Continued)

H_a (cm)	W (cm)								
	15.25	22.86	30.48	45.72	60.96	91.44	121.92	152.40	182.88
65.50			.3628	.5510	.7412	1.126	1.515	1.907	2.302
66.00			.3671	.5575	.7500	1.139	1.533	1.930	2.330
66.50			.3713	.5640	.7588	1.153	1.552	1.953	2.358
67.00			.3755	.5706	.7677	1.167	1.570	1.977	2.386
67.50			.3798	.5771	.7766	1.180	1.588	2.000	2.415
68.00			.3841	.5837	.7855	1.194	1.607	2.024	2.443
68.50			.3884	.5903	.7945	1.208	1.626	2.047	2.472
69.00			.3927	.5970	.8035	1.222	1.645	2.071	2.501
69.50			.3971	.6036	.8125	1.236	1.663	2.095	2.530
70.00			.4014	.6103	.8216	1.249	1.682	2.119	2.559
70.50			.4058	.6170	.8307	1.263	1.701	2.143	2.588
71.00			.4102	.6238	.8399	1.278	1.720	2.167	2.617
71.50			.4146	.6306	.8491	1.292	1.739	2.192	2.647
72.00			.4190	.6373	.8583	1.306	1.759	2.216	2.676
72.50			.4235	.6442	.8675	1.320	1.778	2.240	2.706
73.00			.4279	.6510	.8768	1.334	1.797	2.265	2.736
73.50			.4324	.6579	.8862	1.349	1.817	2.290	2.766
74.00			.4369	.6648	.8955	1.363	1.836	2.314	2.796
74.50			.4414	.6717	.9049	1.378	1.856	2.339	2.826
75.00			.4459	.6787	.9143	1.392	1.876	2.364	2.856

Appendix 3

Standard dimensions of the Parshal flume

W	A	B	C	D	E	F	G	K	N	Rate of flow l/s	
										Min	Max
Small											
25.4 mm	0.242 m	0.356 m	0.093 m	0.167 m	0.229 m	0.076 m	0.203 m	0.019 m	0.029 m	0.283	5.663
76.2 mm	0.306 m	0.457 m	0.178 m	0.259 m	0.457 m	0.152 m	0.305 m	0.025 m	0.057 m	0.849	28.317
Intermediate											
0.152 m	0.415 m	0.610 m	0.394 m	0.397 m	0.610 m	0.305 m	0.610 m	0.076 m	0.114 m	1.416	110.44
0.229 m	0.588 m	0.864 m	0.381 m	0.575 m	0.762 m	0.305 m	0.457 m	0.076 m	0.114 m	2.548	252.02
0.305 m	0.914 m	1.343 m	0.610 m	0.845 m	0.914 m	0.610 m	0.914 m	0.076 m	0.229 m	3.115	455.90
0.457 m	0.965 m	1.419 m	0.762 m	1.025 m	0.914 m	0.610 m	0.914 m	0.076 m	0.229 m	4.247	696.60
0.610 m	1.016 m	1.495 m	0.914 m	1.206 m	0.914 m	0.610 m	0.914 m	0.076 m	0.229 m	11.893	937.29

(Continued)

W	A	B	C	D	E	F	G	K	N	Rate of flow l/s	
										Min	Max
0.914 m	1.118 m	1.645 m	1.219 m	1.572 m	0.914 m	0.610 m	0.914 m	0.076 m	0.229 m	17.273	1427.2
1.219 m	1.219 m	1.794 m	1.524 m	1.937 m	0.914 m	0.610 m	0.914 m	0.076 m	0.229 m	36.812	1922.7
1.524 m	1.321 m	1.943 m	1.829 m	2.302 m	0.914 m	0.610 m	0.914 m	0.076 m	0.229 m	45.307	2423.9
1.829 m	1.422 m	2.093 m	2.134 m	2.667 m	0.914 m	0.610 m	0.914 m	0.076 m	0.229 m	73.624	2930.8
2.438 m	1.626 m	2.391 m	2.743 m	3.397 m	0.914 m	0.610 m	0.914 m	0.076 m	0.229 m	99.109	3950.2
Large											
3.048 m	1.829 m	4.267 m	3.658 m	4.756 m	1.219 m	0.914 m	1.829 m	0.152 m	0.343 m	169.90	5663.4
3.658 m	2.032 m	4.877 m	4.470 m	5.607 m	1.524 m	0.914 m	2.438 m	0.152 m	0.343 m	226.54	9910.9
6.096 m	2.845 m	7.620 m	7.315 m	9.144 m	2.134 m	1.829 m	3.658 m	0.305 m	0.686 m	283.17	28317.0

Bibliography

1. ASTM, "D-1239, Method of Test for Resistance of Plastic Films to Extraction by Chemicals".
2. ASTM, "D-1191, Testing Concrete Joint Sealers".
3. ASTM, "D-5, Test Method for Penetration of Bituminous Material".
4. ASTM, "D-560, Standard Test Methods for Freezing and Thawing Compacted Soil-Cement Mixtures".
5. ASTM, "D-558, Standard Test Methods for Moisture – Density Relations of Soil-Cement Mixtures".
6. ASTM, "C-654, Standard Specifications for Porous Concrete Pipe".
7. ASTM, "D-882, Method of Test for Tensile Properties of Thin Plastic Sheet and Films".
8. ASTM, "D-747, Stiffness of Plastics by means of Cantilever Beam".
9. ASTM, "D-689, Method of Test for Internal Tearing Resistance of Paper".
10. ABBOT M.B., "Computational Hydraulics", Pittmann Publishing Limited, London.
11. ABBOT M.B. and CUNGE J.A., "Engineering Applications of Computational Hydraulics", Water Resources Publication, Fort Collins.
12. ABBOT M.B., "Method of Characteristics", Water Resources Publication, Fort Collins.
13. ALLUÉ ESCUDERO MIGUEL, "Semblanza de D. Ramón Pignatelli. Su vida y su obra". Conferencia dada en la Real Sociedad Económica Aragonesa de Amigos del País, 13 de Abril de 1994, Spanish. ("Portrait of Ramón Pignatelli. His life and His Work". Lecture given in the "Sociedad Aragonesa de Amigos del Pais".)
14. ARAOZ SANCHEZ DE ALBORNOZ ANGEL, "Problemas Geotécnicos en los canales de la cuenca del Ebro. Una larga historia", "Jornadas sobre Canales del Centro de Estudios Hidrográficos", Madrid 1991, Spanish. ("Geotechnical Problems in Canals of the Ebro River Zone. One Long Story". Symposium held in the Hydrografic Centre.)
15. BAKHMETEFF BORIS, "Hidráulica de los Canales", Aguilar, Madrid, Spanish. ("Canal Hydraulic".)
16. BAKHMETEFF BORIS, "Hydraulics of Open Channels", McGraw-Hill, London and New York.
17. BARBANCHO FRANCISCO, "Estudio de un caso real de pérdidas de agua en un canal de riego: Fugas en el Canal deOrellana en el período 1981–1991", "Jornadas sobre Canales del Centro de Estudioa Hidrográficos", Madrid 1991,

Spanish. ("A True Case of Water Losses in an Irrigation Canal. See page in Orellana Canal During the Period 1981–1991". Symposium held in the Hydrografic Studies Centre.)

18. BARBANY ANTONIO, NUÑEZ MAESTRO ANGEL and SANCHO MARCO TOMAS, "Experiencias de reparación de canales con materiales no convencionales", "Jornadas sobre Canales del Centro de Estudioa Hidrográficos", Madrid 1991, Spanish. ("Experiences on Canal Repair with New Materials". Symposium held in the Hydrografic Studies Centre.)

19. BLANCO MANUEL and AGUIAR ESCOLÁSTICO, "Geomembranas sintéticas a base de Poliuretano de alta densidad utilizado en la impermeabilización de embalses. Seguimiento de obras en las Islas Canarias", Revista de Ingeniría Civil, no. 109, Spanish. ("H.Density Polyuretane Geomembranes used for Reservoirs Impermeabilisation – Experiences Got in Canary Islands". Civil Engineering Journal. no. 109.)

20. BOSS MARINUS, "Flow Measurement and Regulating Flumes".

21. BUREAU OF RECLAMATION, "Design of Small Canal Structures".

22. BUREAU OF RECLAMATION, "Paint Manual".

23. BUREAU OF RECLAMATION AND AGRICULTURAL RESEARCH SERVICE, "Chemical Control of Submerged Waterweeds in Western Irrigation and Drainage Canals".

24. BUREAU OF RECLAMATION, "Manual del Hormigón", Editorial Dossat.

25. BUREAU OF RECLAMATION, "Linings for Irrigation Canals".

26. BUREAU OF RECLAMATION, "Canal Systems Automation Manual".

27. BUREAU OF RECLAMATION, "Perfomance of Plastic Canal Linings", 1984.

28. BUREAU OF RECLAMATION, "Water Measurement Manual".

29. BUREAU OF RECLAMATION, "Design of Small Dams".

30. BUREAU OF RECLAMATION, Thomas Haider and Thomas Mitchell, "Informe provisional sintetizado sobre revestimientos de canales utilizados por el Bureau of Reclamation", Spanish. ("Provisional Report on Canal Linings used by the Bureau of Reclamation".)

31. BUREAU OF RECLAMATION, "Earth Manual".

32. BUREAU OF RECLAMATION, "Canal Linings and Methods of Reducing Costs".

33. BUREAU OF RECLAMATION, "Evaluation of Semicircular Canals".

34. BUREAU OF RECLAMATION, "Investigation of Plastic Films for Canal Linings", Research Report no. 19.

35. BUREAU OF RECLAMATION, Chester Joners, "Investigaciones del Bureau of Reclamation sobre instalaciones de trasporte de agua". (Bureau of Reclamation "Investigations on Water Transport Facilities".)

36. CASTAGNY G., "Traité Pratique des Eaux Souterraines", Dunod, Rue Bonaparte, Paris 1971, French. ("Practical Manual on Underground Water".)

37. CENTRAL BOARD OF IRRIGATION AND POWER, "Symposium on Canal Linings", New Delhi 1960.

38. CENTRAL BOARD OF IRRIGATION AND POWER, "Manual on Canal Linings", New Delhi, September 1975 (reprinted March 1979).

39. CENTRO DE ESTUDIOS HIDROGRÁFICOS, "Recomendaciones para el Proyecto de Canales", Ministerio de Obras Públicas. ("Recomendations on Canal Design", Ministry of Public Works of Spain.)

40. CUNGE JEAN, "Applied Mathematical Modelling of open Channel Flow", Water Resources Publication, Fort Collins, Colorado 1975, USA.
41. DEL CAMPO JOAQUÍN, "Canal del Genil-Cabra", "Jornadas sobre Canales del Centro de Estudios Hidrográficos", Madrid 1991, Spanish. ("Genil-Cabra Canal", Symposium held at the Hydrografic Centre.)
42. DIN 19.704, "Hydraulic Steel Structures Criteria for Design and Calculation".
43. DIN 19.705, "Hydraulic Steel Structures Recommended for Design, Construction and Erection".
44. DOMINGUEZ, "Hidráulica", Universidad Católica de Santiago de Chile, Spanish. ("Hydraulics" Catholic University, Santiago de Chile.)
45. FAO, "Small Hydraulic Structures".
46. GOMEZ NAVARRO and ARACIL J.L., "Saltos de agua y Presas de Embalse", Escuela de Ingenieros de Caminos, Canales y Puertos, Madrid, Spanish. ("Hydroelectric Power and Regulating Dams", Roads, Canals and Harbours College of Madrid.)
47. GOUSSARD JEAN, "Automation of Canal Irrigation Systems", ICID.
48. GUISADO FRANCISCO, "Costo de los Canales a cielo abierto", "Jornadas sobre Canales del Centro de Estudioa Hidrográficos", Madrid 1991, Spanish. ("Open Canal Cost", Symposium held at the Hydrografic Centre. Madrid.)
49. HENDERSON F.M., "Open Channel Flow", McGraw-Hill Book Company, London and New York.
50. ENOCQUE A., "Les revętements de canaux en béton bitumineux répandu en place", TRAVAUX, no. 313, Edition Science et Industrie, November 1960, French. ("Hot Mix Canal Lining Constructed on Site".)
51. HICKERY M.E., "Synthetic Rubber Lining", Bureau of Reclamation.
52. HOLTH W.G., "Thick Compacted Earth Linings for Canals", 3rd ICID Congress Congreso de la ICID.
53. HUÉ HERRERO FERNANDO, "Canal Problems in Gypserous Soils", 3rd ICID Congress.
54. INTERNATIONAL COMMISSION ON IRRIGATION AND DRAINAGE, "First Draft on Aquatic Weed Control in Canals and Reservoirs", 1999.
55. INTERNATIONAL COMMISSION ON IRRIGATION AND DRAINAGE, "Design Practices of Irrigation Canals", 1972.
56. INTERNATIONAL COMMISSION ON IRRIGATION AND DRAINAGE, "Controlling Seepage Losses in Irrigation Canals, World Wide Survey", 1973.
57. KAZEN SIHAI M., HASSAN RAHIMI, AHMAD JAFARI, MADDAH M., GHOBBI SH. and JAVAD MOULAIE, "Rehabilitation of Irrigation Canals in IRAN", ICID Journal no. 2, 1997.
58. KAZIMIERZ GARBULEWSKI, "Damage to Irrigation Enmankment Canals Constructed with Expansive Soils", 15th Congress on Irrigation and Darainage ICID".
59. KHASIN A., "Investigation of Sealing Materials for Making Watertight Slab Juntions and Expansion Joints of Reinforced Concrete Canal Linings", Traducción del ruso-Gidrotekniche Skoe Stroitel Stvo, translated from Russian, January 1969.
60. LELIAVSKY, "Précis d'Hydraulique Fluviale Apliquée", French. ("Applied Essential Fluvial Hydraulics", Dunod, Paris.)
61. LIGGET JAMES, "Stability", Water Resources Publication, Fort Collins.

62. LIGGET J. and CUNGE J., "Numerical Methods of Solution of the Unsteady Flow Equations", Water Resources Publication, Fort Collins 71.
63. LIGGET JAMES, "Basic Equations of Unsteady Flow", Water Resources Publication, Fort Collins.
64. LINSLEY, KOEHLER and PAULUS, "Hidrología para In genieros", Ediciones Castillo. Spanish. ("Hydrology for Engineers", Castillo Publisher.)
65. LIRIA MONTAÑÉS JOSÉ and TORRES PADILLA CARLOS, "Un ejemplo de acequias prefabricadas autoportantes", Centro de Estudios Hidrográficos, Spanish. ("One Example of Elevated Terciary Canals", Hydrografic Studies Center.)
66. LIRIA MONTAÑÉS JOSÉ, "Proyecto de Redes de Distribución de Agua", Colegio de Ingenieros de Caminos. Colección Seinor 1995, Spanish. ("Design of Pipe Networks for Water Supply", Civil Engineering Institution, Madrid, Seinor Collection 1995.)
67. LIRIA MONTAÑÉS JOSÉ, "Posibilidades de ahorro de Agua en España mediante una acertada modulación y explotación de los canales de riego", Primer Simposio Nacional Presente y Futuro de los Regadíos Españoles, Madrid 1994, Spanish. ("Possibilities of Saving Water in Spain by Means of an Adequate Management of Irrigation Canals", First National Congress on "Present and Future of the Spanish Irrigated Zones", Madrid 1954.)
68. LIRIA MONTAÑÉS JOSÉ, "A Mathematical Model Helps to Avoid Spilled Water Losses in Canals", Congreso Internacional de ICID, Granada 1999.
69. LIRIA MONTAÑÉS JOSÉ "Modelo Canvar. Simulación del Movimiento Variable del Agua en un Canal de Riego", Revista de Ingeniería Civil no. 104, Spanish. ("Canvar Model for Simulation of Unsteady Flow in an Irrigation Canal", Civil Engineering Journal no. 104.)
70. LIRIA MONTAÑÉS JOSÉ, "Fonctionnement Hydraulique des canaux de Trasvasement. Etude d'un cas particulier", XIV Journées Regionales Européennes, French. ("Hydraulic Operation of Water Transfer Canals. An Example", XIV European Meeting on Irrigation and Drainage, May 1986.)
71. LIRIA MONTAÑÉS JOSÉ, "Programa Canvar. Un modelo matemático para simular el funcionamiento hidráulico de un Canal", Revista de Riegos y Drenajes XXI Mayo de 1996, Spanish. ("Canvar Programme. A Mathematical Model for Simulation of Hydraulic Canal Operation", Irrigation and Drainage XXI Journal, May 1996.)
72. LIRIA MONTAÑÉS, Spanish. ("Small Structures for Terciary Irrigation Canals", Hidrographic Studies Centre.)
73. LOUBATON, "Les revetments des canaux en béton de ciment", TRAVAUX, Noviembre 1960, French. ("Concrete Canal Linings" Travaux, November 1960.)
74. MACAU VILLAR, "Estabilización de taludes en desmontes y terraplenes", Boletín no. 8 del Servicio Geológico de Obras Públicas, Spanish. ("Slope Stabilisation on Excavations and Backfillings", Bulletin no. 8 of the Geological Service for Public Works.)
75. MAHMOOD K. and YEVJEVICH V., "Unsteady Flow in Open Channels", Water Resources Publication, Fort Collins, 1975.
76. MANZANARES J.L. and SAURA J.L., "Comparación entre un canal trapecial y otro circular en terrenos expansivos", Revista de Obras Públicas, Octubre

1985, Spanish. ("Comparison of a Circular Canal with a Trapezoidal One in Expansive Ground", Public Works Magazine.)

77. MINISTERIO DE FOMENTO, "Instrucción de Hormigón Estructural", EHE, Spanish. ("Standards for Structural Concrete".)

78. MOLIN M., "Revêtement des canaux principaux au Maroc", 3rd Congreso de Riegos y Drenajes ICID, French. ("Canal Linings in Morocco", International Commission on Irrigation and Drainage, 3rd Congress, San Francisco 1957.)

79. ORTEGA GÓMEZ EDUARDO-PROSER, "Simulación de la Explotación-Estudio en Régimen Variable, programa de Ordenador y Bases de Cálculo", "Jornadas sobre Canales del Centro de Estudioa Hidrográficos", Madrid 1991, Spanish. ("Canal Management Simulation. Unsteady Flow. Computer Program and Calculation Basis", Symposium on Canals, Hidrographic Studies Centre, Madrid 1991.)

80. ORTEGA JUAN, "Juntas en pavimentosde hormigón", "Laboratorio Central de Ensayo de Materiales de construcción", Madrid, Spanish. ("Joints for Concrete Pavements", Central Laboratory for Construction Materials, Madrid 1957.)

81. PLUSQUELECQ HERVÉE, "Improving the Operation of Canal Irrigation Systems", Bureau of Reclamation.

82. RAHIMI H. and ABBASI N., "Soil Cement Tiles for Lining of Irrigation Canals", ICID Journal, Vol 48, August 1999.

83. SANTOS GARCÍA FLORENTINO, "Eficiencia de los Canales de Riego", "Jornadas sobre Canales del Centro de Estudios Hidrográficos", Madrid 1991, Spanish. ("Irrigation Canals Efficiency", Symposium on Canals, Hidrographic Studies Centre, Madrid 1991.)

84. SAURA MARTINEZ JUAN, "Influencia de las arcillas expansivas sobre la elección de la sección trasversal en canales", "Jornadas sobre Canales del Centro de Estudioa Hidrográficos", Madrid 1991, Spanish. ("Expansive Clay Influence on Canal Cross Section. Symposium on Canals". Hidrographic Studies Centre, Madrid 1991.)

85. SOIL CONSERVATION SERVICE, "Measuring Water in Irrigation Channels with Parshall Flumes and Small Weirs".

86. SOUBRIER GONZALO, "Pérdidas de Agua y problemas relacionados con ella. Drenaje", "Jornadas sobre Canales del Centro de Estudioa Hidrográficos", Madrid 1991, Spanish. ("Water Losses and Related Problems. Drainage", "Symposium on Canals", Hidrographic Studies Centre, Madrid 1991.)

87. SOUBRIER GONZALO, "Canal de las Dehesas. Riego y Trasvase", Revista Cauce 2000, Noviembre y Diciembre 1988, Spanish. ("Dehesas Canal. Irrigation and Transfer", Cauce 2000 Magazine, November and December 1988.)

88. STREETER VICTOR and WYLIE BENJAMIN, "Hydraulic Transients", McGraw-Hill.

89. STREK OTTO, "Problemas de Hidraulica aplicada", Editorial Labor. Madrid Buenos Aires Rio de Janeiro 1948.

90. SZECHY KAROLY, "The Art of Tunnelling", Akadémiai Kiadó, Budapest 1970.

91. TEMEZ JOSÉ RAMÓN, "Cálculo hidrometeorológico de caudales máximos en pequeñas cuencas naturales", Dirección General de Carreteras Mopu 1978, Spanish. ("Hidrometeorological Calculation of Maximum Flows in Smalls Natural Bassins", Road Directorate Mopu 1978.)

92. TERZAGHY Y PECK, "Soil Mechanics in Engineering Practice", Wiley, London 1967.
93. TORRENT LUIS, "Resaltos en canales", Revista de Obras Públicas, Octubre de 1993, Spanish. ("Hydraulic Jump in Canals", Public Works Journal, October 1993.)
94. USSR NATIONAL COMMiTTEE ON IRRIGATION AND DRAINAGE, "Manual on Designing and Construction of Irrigation Canals", Moscow 1977.
95. VALIRON F., "Problèmes posés pour le revêtement des canaux par suite de la nature de terres", 3º Congreso de la ICID, French. ("Canal Lining Problems Derived from Nature of the Land", 3rd Congress of ICID.)
96. VAN ASBECK, "Le bitume dans les travaux hydrauliques", Dunod, French 1962. ("Bitumen in Hydraulic Works".)
97. VEN-TE-CHOW, "Handbook of Applied Hydrology", McGraw-Hill, London and New York 1964.
98. VEN-TE-CHOW, "Open Channel Hydraulics", McGraw-Hill, London and New York 1982.

Index

Also available from Taylor & Francis

Hydraulics in Civil and Environmental Engineering

Andrew Chadwick, John Morfett, Martin Borthwick

| | Spon Press | Pb: 0-415-30609-4 |

Sediment Transport in Irrigation Canals

V. Mendez, J. Nestor

| | A A Balkema | Hb: 90-5410-413-9 |

Fluvial, Environmental and Coastal Developments in Hydraulic Engineering

Edited by: Michele Mossa, Youichi Yasuda, Hubert Chanson

| | Spon Press | Hb: 0-415-35899-X |

Hydraulic Structures

P. Novak, A.I.B. Moffat, C. Nalluri, R. Narayanan

| | Spon Press | Hb: 0-415-25070-6 |

Information and ordering details

For price availability and ordering visit our website **www.sponpress.com**
Alternatively our books are available from all good bookshops.